Do**Brilliantly**

ASChemistry

Exam practice at its **best**

■ **George Facer**

■ **Series Editor: Jayne de Courcy**

I would like to dedicate this book to my father, John Facer, who taught me and inspired me with a lifelong love of Chemistry.

Published by HarperCollins*Publishers* Limited
77–85 Fulham Palace Road
London W6 8JB

www.**Collins**Education.com
On-line support for schools and colleges

First published 2001
10 9 8 7 6 5 4 3
ISBN 0 00 710705 6

British Library Cataloguing in Publication Data
A catalogue record for this book is available from the British Library

Edited by Kathryn Senior
Production by Kathryn Botterill
Cover design by Susi Martin-Taylor
Book design by Gecko Limited
Printed and bound in China by Imago

Acknowledgements
Chris Conoley for reading and commenting on the manuscript.

Illustrations
Cartoon Artwork – Roger Penwill
DTP Artwork – Richard Morris

Contents

How this book will help you by George Facer

Exam practice – how to answer questions better

This book will help you improve your performance in your AS Chemistry exam. It contains lots of **questions on the core topics of the new AS specifications**.

In exams, students often fail to gain the grades they are capable of, even when they have worked hard throughout their course and have a sound grasp of the facts. It is often not a lack of knowledge that leads to disappointing results, but poor examination technique. As an examiner, I often wish that I could sit alongside students as they answer questions, so that I could prompt them to make their explanation a bit fuller or ask them to clarify a point. **This book will help you with your examination technique so that you can make the most of your knowledge and score high marks in your exam**.

Each chapter in this book is broken down into four separate elements, aimed at giving you as much guidance and practice as possible:

1 Exam Question, Student's Answer and 'How to score full marks'

The questions and students' answers that I have chosen to start each chapter are typical ones. They show a number of mistakes frequently made by candidates under exam conditions.

The 'How to score full marks' section explains precisely where the student lost marks, e.g. not following instructions in the question, not putting in enough detail, etc. **I show you how to pick up those vital extra marks that make all the difference between an ordinary grade and a very good one**.

2 'Don't make these mistakes'

This section highlights the most common mistakes students make either in the exam itself or in their preparation for the exam. When you are into your last minute revision, you can quickly read through all of these sections and make doubly sure that you avoid these mistakes in your exam.

3 'Key points to remember'

These pages list **some of the most important facts you need to know or definitions you need to learn for each topic**. They are not meant to replace your notes, but are a quick check on those points that it is vital you know before going into your exam. You'll find these pages really helpful with last minute revision.

4 Questions to try, Answers and Examiner's comments

Each chapter ends with a number of exam questions for you to answer. Don't cheat. Sit down and answer the questions as if you were in an exam. Try to put into practice all that you have learnt from the previous sections in the chapter. I've included some exam hints before each question which should help you get the correct answers. Check your answers through and then look at the answers given at the back of the book. These are full mark answers.

In the 'Examiner's comments', I highlight anything tricky about the question which may have meant you did not get the correct answer. By reading through these sections, you can avoid making mistakes in your actual exam.

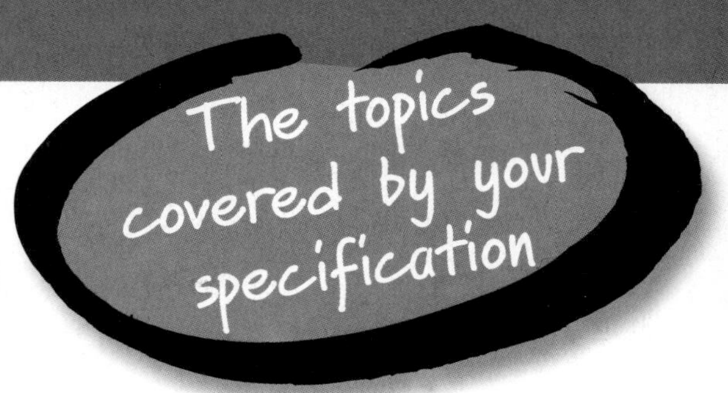

ll exam boards have the same core of topics in AS hemistry, but outside this core each has freedom bout what they may put in their specification syllabus). In this book I have concentrated on including **uestions on the areas that are common to all oards**. I have also, in some cases, added topics which re in the specification of two of them. Terminology aries from exam board to exam board and I have tried o take this into account as far as possible.

he chart below lets you see at a glance which chapters n this book are relevant to your particular specification.

	AQA	EDEXCEL	NUFFIELD	OCR	WJEC
hapter 1 tomic structure	Unit 1	Unit 1	Unit 1	Unit 2811	CH1
hapter 2 ormulae, quations and mount of ubstance	Unit 1	Unit 1 and Unit 3B	Units 1 & 2	Unit 2811	CH1 and CH3a
hapter 3 tructure and onding	Unit 1	Unit 1	Ionic & metallic Unit1 Covalent Unit 2	Unit 2811	CH1
hapter 4 eriodic Table & roup II	Unit 1	Unit 1	Unit 1	Unit 2811	CH1
hapter 5 edox & Group VII	Unit 2	Unit 1	Unit 2	Unit 2811	CH1
hapter 6 nergetics	Unit 2	Unit 2	Unit 1	Unit 2813/1	CH2
hapter 7 inetics	Unit 2	Unit 2	Unit 2	Unit 2813/1	CH2
hapter 8 quilibrium	Unit 2	Unit 2	Unit 2	Unit 2813/1	CH2
hapter 9 rganic Chemistry	Unit 3	Unit 2	Alcohols Unit 1 Hydrocarbons & halogenoalkanes Unit 2	Unit 2812	CH2

Note: AQA also includes the extraction of iron and aluminium in unit 2 nd the distillation of petroleum and the cracking of a fraction in unit . Edexcel includes industrial chemistry (the manufacture of ammonia, lphuric acid, aluminium and chlorine) in unit 2. OCR includes the racking, isomerisation and reforming of fractions of crude oil in unit 812. These topics are not covered in this book due to lack of space.)

- **Know the rubric**. Do you have to answer all the questions? How many questions are there on the paper (this may vary from year to year)? How much time do you have for each question? As a rough guide the rate is about 1 mark per minute.
- If a question is broken down into sections, **consider each section as part of the whole**, because the various parts will be linked.
- Read each section of a question carefully and fully before answering it. **Highlight any command words** (see below).
- The total marks for each part will be shown in or near the margin. **If there are two marks for a part, make sure that you write at least two statements that are worthy of scoring one point each**.
- **Set out calculations clearly**, showing what you are calculating at each step. Give your final answer to the correct number of significant figures (if in doubt give it to 3) and add the unit. Don't round up intermediate answers to 1 or 2 significant figures.
- **If you want to alter something, don't use Tippex**. Neatly cross it out and write your answer in a free space in the question paper. If it is on a different page, alert the examiner by writing 'see page xx'. If you change your mind and want the examiner to mark what you originally wrote, write 'please ignore crossing out'.
- If you are asked for a reagent or name, **don't give alternatives**. If one is wrong, you will be penalised.
- When you have finished, **check as many of your answers as you have time for**:
  - **Start with a calculation**. Have you worked out the relative molecular mass correctly? Have you misread any numbers? Re-calculate the answer to check that you have not made a calculator error. Finally check the number of significant figures and the units of your answer.
  - **Check all chemical formulae**. Make sure that you have not made any silly errors such as $KMnO_4^-$ (this is surprisingly common). Check the oxidation state in names.
  - **Check that all equations balance**.
  - **Beware of contradictory statements** such as conc H_2SO_4(aq) or NaOH in acid.
  - Check that you have given **the full name or formula** for a reagent.
  - In organic chemistry **check formulae** to ensure that all carbon atoms have 4 bonds, all oxygen atoms 2 and hydrogen only 1.
- **Make sure you really understand the meaning of these 'command words'**:
  - **Define** – This means that you must recall the formal textbook definition. It is often helpful to include an example or equation to supplement your definition.
  - **Explain** – This requires a detailed answer. Look to see how many marks there are, and make at least the same number of points.
  - **State** – A simple, often one-word answer all that is required. There is no need for any explanation.
  - **Deduce** – Use the data supplied to answer the question.
  - **Hence deduce** – Use the answer that you have just obtained in the previous part to work out the answer to this part.
  - **Suggest** – You are not expected to have learnt the answer. You should be able to apply your knowledge of similar reactions or substances and work out the answer.
  - **Compare** – You must say something about *both* substances or reactions.
  - **Name** – Give the full name not the formula.
  - **Structural formula** – It is safest to give a formula showing all the atoms and all the bonds. Never use added up formula when it is ambiguous, e.g. for the C_3H_7 group.

1 Atomic structure

Exam Question and Student's Answer

1 (a) What is the electronic configuration of a Mg atom and a Mg^{2+} ion?

Mg is $1s^2, 2s^2\ 2p^6, 3s^2$ ✓

Mg^{2+} is $1s^2, 2s^2\ 2p^6$ ✓

[2 marks] 2/2

(b) (i) Define the term second ionisation energy with reference to magnesium. Illustrate your answer with an equation.

The second ionisation energy is the energy required ∧ to remove two ✗ electrons.

$Mg(g) \rightarrow Mg^{2+}(g) + 2e^-$ ✗ ✓ ✗

[3 marks]

(ii) Why are ionisation energies always endothermic?

It is because the negative electron ✓ is attracted to the positive nucleus. ∧

[2 marks]

(iii) Explain why the third ionisation energy of magnesium is very much larger than the second.

The 3rd electron comes from the 2nd shell ✓ and this shell is closer to the nucleus ✓ and so the 2nd shell electrons are held more firmly. ∧

[3 marks] 2/3

(c) Magnesium exists as isotopes. Define the term isotopes, explaining in what way they differ.

An isotope is a different atom from the atom of the other isotope.

They have a different number of neutrons.

[4 marks] 2/4

(d) The variation of first ionisation energies with atomic number for the first ten elements is shown below.

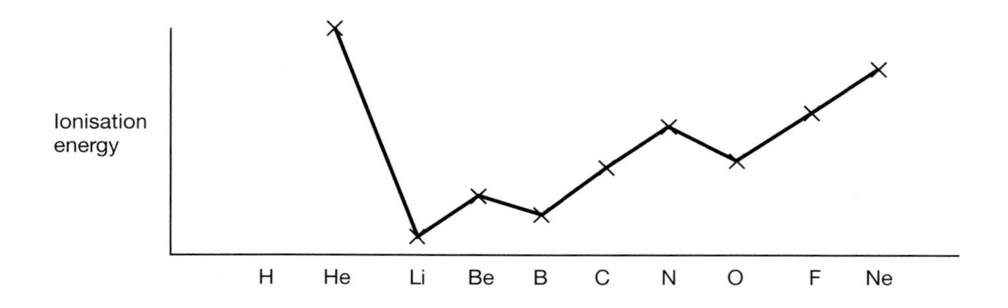

(i) Explain why helium has the largest ionisation energy.

Helium is by far the smallest atom of all in the Periodic Table, and so the force of attraction between the nucleus and the outer unshielded, 1s, electron is the largest of all.

[2 marks] 

(ii) Explain why the first ionisation energy of beryllium is greater than that of lithium and also is greater than that of boron.

The nuclear charge in a beryllium atom is 1 more than that of lithium, and so more energy is required to remove the 2s electron in beryllium than the 2s electron in lithium. In boron the electron removed is in the higher energy 2p orbital and so is easier to remove.

[4 marks] 3/4

[Total 20 marks]

How to score full marks

b) (i) The answer is wrong. The 2nd ionisation energy is **the energy per mole required to go from $Mg^{+}(g)$ to $Mg^{2+}(g)$, not** from Mg(g) to $Mg^{2+}(g)$. The student also fails to mention that the magnesium ions must be **gaseous**, but the equation with state symbols shows this and so scores the mark. The correct answer is that the 2nd ionisation energy is the energy required to remove one electron from **one mole** of **gaseous Mg^{+}** ions, i.e. for the energy change per mole for the process:

$$Mg^{+}(g) \rightarrow Mg^{2+}(g) + e^{-}$$

(ii) To score the second mark you must make it clear that the process of **overcoming** the force of attraction between the electron and the nucleus **requires** energy and so is endothermic.

(iii) To gain the third mark, you must state that the electrons in the 2nd shell are **shielded only by the two** electrons in the 1st shell.

c) **There are four marks for this part of the question and you need to make all four points to gain them all**. The student only made two points and so only scored two marks. You would gain the third mark for emphasising that isotopes are different atoms of the **same element**. You would gain the fourth mark for stating that the different number of neutrons results in the isotopes having different **mass numbers**.

d) (ii) A mark was lost for failing to state that the **shielding** by the inner electrons is the **same** for both beryllium (Be) and lithium (Li).

Don't make these mistakes ...

In your definition of an s-block element, do **not** state that 'an s-block element has a **partially** filled s orbital'. Beryllium is an s-block element but its s orbital is full; its electronic configuration is $1s^2 2s^2$. The reason that beryllium is an s-block element is that its **highest energy** electron is in an s orbital. (The same rule holds true for deciding whether an element is a *p*- or *d*-block element.)

Do not forget to give an **equation, including state symbols**, in your definition of ionisation energy and electron affinity. This may help you to gain full marks if you leave out a scoring point such as the fact that the atoms must be gaseous.

Mass spectra are caused by **positive** ions. Do **not** leave out the '+ sign' in the formula of an ion that causes a specific line in a mass spectrum.

Ionisation energies are always related to the **removal** of an electron. For chlorine the process

$$Cl(g) + e^{-} \rightarrow Cl^{-}(g)$$

is called the (first) **electron affinity** of chlorine.

In your explanation of isotopes do **not** imply that they are different **elements**, but stress that they are different **atoms** of the **same** element.

Key points to remember

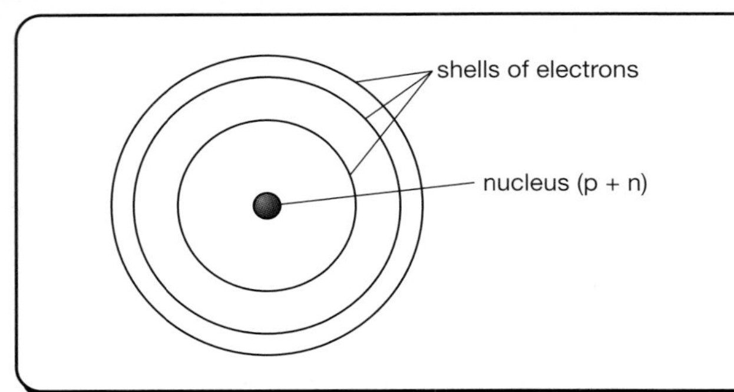

Atomic (or proton) number (Z) is the number of protons in the nucleus.

Mass number (A) of an isotope is the sum of the number of protons and neutrons in the nucleus.

Orbit(shell)	orbitals	max no of electrons
1	s	2
2	s & p	2 + 6 = 8
3	s, p & d	2 + 6 + 10 = 18

The order of filling orbitals is

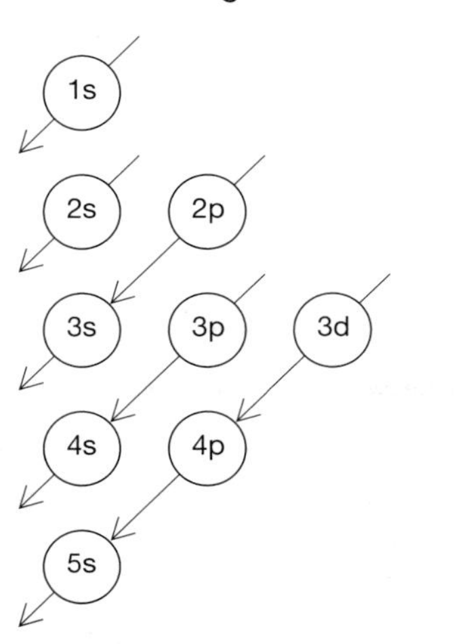

Electron structures are:

$_{8}O$	$1s^2, 2s^2\ 2p^4$	
$_{10}Ne$	$1s^2, 2s^2\ 2p^6$	
$_{17}Cl$	$1s^2, 2s^2\ 2p^6, 3s^2\ 3p^5$	or $[Ne]\ 3s^2, 3p^5$
$_{18}Ar$	$1s^2, 2s^2\ 2p^6, 3s^2\ 3p^6$	
$_{20}Ca$	$1s^2, 2s^2\ 2p^6, 3s^2\ 3p^6, 4s^2$	or $[Ar]\ 4s^2$
$_{26}Fe$	$1s^2, 2s^2\ 2p^6, 3s^2\ 3p^6\ 3d^6, 4s^2$	or $[Ar]\ 3d^6, 4s^2$

Mg is *s*-block because its electronic structure is $1s^2\ 2s^2\ 2p^6\ 3s^2$

Ar is *p*-block because its electronic structure is $1s^2\ 2s^2\ 2p^6\ 3s^2\ 3p^6$

Ti is *d*-block because its electronic structure is $1s^2\ 2s^2\ 2p^6\ 3s^2\ 3p^6\ 3d^2\ 4s^2$ (3d is of higher energy than 4s).

An ***s*-block element** is one where the highest occupied energy level is an s orbital.

A ***p*-block element** is one where the highest occupied energy level is a p orbital.

A ***d*-block element** is one where the highest occupied energy level is a d orbital.

Carbon has three naturally occurring isotopes: ^{12}C, ^{13}C and ^{14}C

Mass number / Atomic number	$^{12}_{6}C$	$^{13}_{6}C$	$^{14}_{6}C$
No of protons	6	6	6
No of neutrons	6	7	8

Isotopes are different atoms of the same element. The have the same number of protons, but a different number of neutrons. Thus they have different masses.

The **relative atomic mass** of an element is the average of the masses of the naturally occurring isotopes (taking into account their abundances) divided by $\frac{1}{12}$th the mass of a carbon 12 atom.

Chlorine has two isotopes:

^{35}Cl with an abundance of 75%, and

^{37}Cl with an abundance of 25%.

The average mass is $\frac{(35 \times 75 + 37 \times 25)}{100} = 35.5$

Mass spectra are caused by a gaseous substance being bombarded by high energy particles (electrons). The substance forms positive ions that are accelerated by an electric field and then deflected by a magnetic field. The stream of ions is split up according to the $\frac{m}{z}$ (mass/charge) value of the ions. The ions produced may be molecular ions or fragments (so bromine will have two lines at $\frac{m}{z}$ of 79 and 81 as well as those shown in Fig 1.4).

The **1st ionisation energy** is the energy required to remove one electron from one mole of gaseous atoms.

The 2nd ionisation energy is the energy required to remove one electron from one mole of gaseous ions with a single positive charge.

The value of the 1st ionisation energy depends mainly upon the pull of the nucleus on the electron that is removed (the stronger the pull, the larger the ionisation energy). This pull is determined by two factors:

- The effective nuclear charge (this equals the atomic number minus the number of inner or shielding electrons);
- The distance from the nucleus to the electron (the smaller the atom, the larger the ionisation energy).

The **mass spectrum of molecular bromine is**:

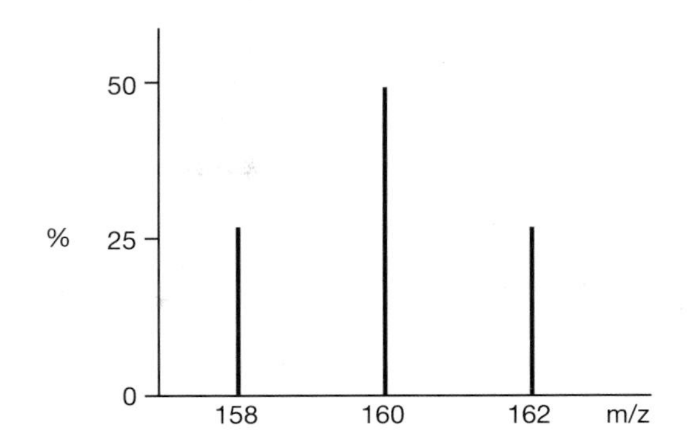

$\frac{m}{z}$ = 158 is caused by $(^{79}Br\ ^{79}Br)^+$

$\frac{m}{z}$ = 160 is caused by $(^{79}Br\ ^{81}Br)^+$

$\frac{m}{z}$ = 162 is caused by $(^{81}Br\ ^{81}Br)^+$

As the lines at $\frac{m}{z}$ of 158 and 162 are of equal height, there is 50% of each isotope.

The **1st ionisation energy** of chlorine is the energy per mole for

$$Cl(g) \rightarrow Cl^+(g) + e^-$$

The **2nd ionisation energy** of chlorine is the energy per mole for

$$Cl^+(g) \rightarrow Cl^{2+}(g) + e^-$$

Ionisation energies are always positive (endothermic).

Across a period the **atomic number increases** but there is no increase in shielding since the number of inner electrons remains the same. Also, the atoms get slightly smaller across a period. These factors cause a general increase in the ionisation energy across a period.

Down a group, the atomic number increases, but so does the number of shielding electrons. Thus, the effective nuclear charge remains the same, but the outer electron becomes further from the nucleus and so is easier to remove.

The 1st electron affinity for chlorine is the energy per mole for:

$$Cl(g) + e^- \rightarrow Cl^-(g)$$

1st electron affinities are always **negative** (exothermic).

The 1st electron affinity is the energy change when 1 electron is added to each of one mole of gaseous atoms.

Questions to try

Examiner's hints ***for question 1***

(a) Elements are classified into blocks according to the highest occupied energy level.
Ionisation energies refer to the **loss** of electrons. Their values depend on the strength of the force of attraction between the nucleus and the electron being removed. This is determined by:

- the charge on the nucleus;
- to what extent the nucleus is shielded by inner electrons;
- the distance between the nucleus and the electron being removed;
- any repulsion between two electrons in a given orbital.

Q1

Antimony (Sb) and nitrogen (N) are both group 5 elements.

(a) (i) Write the electronic structures, using the s, p, d notation, of nitrogen.

....

[1 mark]

(ii) Why is nitrogen classified as a *p*-block element?

....

[1 mark]

(iii) Write the equation for the first ionisation energy of nitrogen.

....

....

....

[2 marks]

(iv) Why is the first ionisation energy of nitrogen more than that of oxygen but less than that of fluorine?

....

....

....

....

....

[4 marks]

(b) Antimony has isotopes of mass numbers 121 and 123, and a relative atomic mass of 121.8.

(i) Distinguish between the terms **mass number** and **relative atomic mass**.

Mass number ....

....

....

Relative atomic mass ....

....

....

[5 marks]

(ii) The mass spectrum of antimony is shown below. Use the data to show why the relative atomic mass of antimony is 121.8.

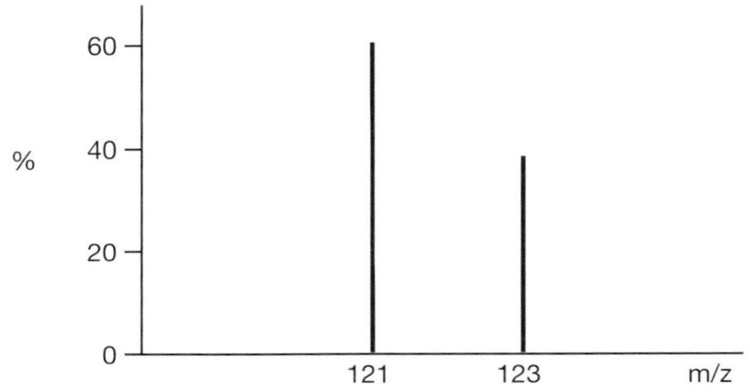

[2 marks]

(iii) In a mass spectrometer the atoms of an element are bombarded by high energy electrons and form positive ions. Describe how these ions are subsequently separated according to their mass/charge $\left(\frac{m}{z}\right)$ ratio.

....

....

....

[3 marks]

[Total 18 marks]

Examiner's hints ***for question 2***

(a) In a neutral atom the number of protons equals the number of electrons. Isotopes are different atoms of the same element.

(b) See **(a)** in examiner's hints for question 1.

Q2

Study the data about the five atoms or ions labelled A to E and answer the following questions.

	A	B	C	D	E
number of protons	11	12	12	12	12
number of neutrons	13	12	13	14	14
number of electrons	11	10	12	13	12

(i) Which are isotopes? ....

[1 mark]

(ii) Which is a positive ion? ....

[1 mark]

(iii) Which is a negative ion? ....

[1 mark]

(iv) Which is a different element from all the others? ....

[1 mark]

(v) Which are atoms? ....

[1 mark]

(b) All the data above refers to elements in the s-block and in period 3 of the Periodic Table.

(i) Explain why the 1st ionisation energies decrease in group 2 from beryllium to barium.

....

....

....

.... [4 marks]

(ii) Explain why the 1st ionisation energies in period 3 increase aluminium to silicon to phosphorus.

....

....

.... [3 marks]

The answers to these questions are on pages 77–78. [Total 12 marks]

2 Formulae, equations and amount of substance

Relative atomic masses required for questions in this chapter are:
[H = 1, C = 12, O = 16, Na = 23, P = 31, Cl = 35.5 and Fe = 56]

Exam Questions and Student's Answers

1 (a) Haematite, Fe_2O_3, is a major ore of iron. In the blast furnace most of the ore is reduced to iron by reduction with carbon monoxide at a high temperature.

(i) Balance the equation below for the reduction of haematite:

$Fe_2O_3(s)$ + $CO(g)$ → $Fe(l)$ + $CO_2(g)$

$Fe_2O_3(s)$ + $3CO(g)$ → ✗ $Fe(l)$ + $3CO_2(g)$

[1 mark] 0/1

(ii) Calculate the volume of carbon monoxide required to reduce 1.80 kg of Fe_2O_3. (1 mole of gas occupies 30.0 dm^3 under these conditions.)

mass of Fe_2O_3 = 1.80 X 1000 = 1800 g ✓

molar mass of Fe_2O_3 = (2 X 56 + 3x 16) = 160 g mol^{-1}

amount of Fe_2O_3 = $\frac{1800}{160}$ = 11.25 mol ✓

amount of CO required = 11.25 X 3 = 33.75 mol ✓

volume of CO required = 33.75 X 30.0 = 1012.5 dm^3 ✗

[4 marks] 3/4

(iii) Calculate the number of iron **atoms** in 1.80 kg of Fe_2O_3. (The Avogadro constant is 6.02×10^{23} mol^{-1}.)

Number of atoms = 11.25 ✓X 6.02 X 10^{23} ∧

= 6.77 X 10^{24} atoms ✗

[2 marks] 

(b) Iron(III)oxide, Fe_2O_3, reacts with sulphuric acid to form iron(III)sulphate and water. Write a balanced equation for this reaction.

Fe_2O_3 + 2✗ H_2SO_4 → 2 $FeSO_4$ ✗ + 2 H_2O

[2 marks] 

(c) One method of preventing rusting iron from getting worse is to treat it with phosphoric acid. This reacts to form an insoluble iron phosphate, which acts as a physical barrier preventing oxygen and water from getting into contact with the iron. A sample of the iron phosphate was analysed and was found to contain 37.1% iron, 20.5% phosphorus and 42.4% oxygen. Calculate the empirical formula of the iron phosphate.

	%	$\% \div A_r$ ✓	$\div 0.66$ ✓
iron	37.1	$\frac{37.1}{56} = 0.66$	1
phosphorus	20.5	$\frac{20.5}{31} = 0.66$	1
oxygen	42.4	$\frac{42.4}{16} = 2.65$	4

Empirical formula is $FePO_4$ ✓

[3 marks] 3/3

7/12

2 (a) Calculate the number of moles of:

(i) hydrogen in 68 cm^3 of H_2 gas at room temperature and pressure. (1 mol of gas occupies 24 dm^3 at room temperature and pressure.)

moles $H_2 = \frac{68}{24}$ ✗ $= 2.8$ mol

[1 mark] 0/1

(ii) sodium hydroxide in 33 g NaOH.

moles NaOH $= \frac{33}{40} = 0.83$ mol ✓

[1 mark] 1/1

(iii) sodium hydroxide in 27.3 cm^3 of 0.123 mol dm^{-3} sodium hydroxide solution.

moles NaOH $= \frac{27.3}{1000} \times 0.123 = 3.36 \times 10^{-3}$ ✓

[1 mark] 1/1

(b) Calculate the mass of $Na_2CO_3.10H_2O$ crystals needed to make 250 cm^3 of a 0.164 mol dm^{-3} solution of sodium carbonate.

250 cm^3 is $\frac{1}{4}$ of 1 dm^3

moles required $= \frac{1}{4} \times 0.164 = 0.0410$ ✓

mass = moles $\times M_r$ ✓ $= 0.0410 \times 106$ ✗ $= 4.35$ g ✗

[3 marks] 2/3

(c) In a titration $25.0\,cm^3$ of the 0.164 mol dm^{-3} solution of sodium carbonate required $28.7\,cm^3$ of a solution of hydrochloric acid. They react according to the equation:

$$Na_2CO_3(aq) + 2HCl(aq) \rightarrow 2NaCl(aq) + CO_2(g) + H_2O(l)$$

Calculate the concentration of the hydrochloric acid solution:

(i) in mol dm^{-3}
(ii) in g dm^{-3}

(i) volume of $Na_2CO_3(aq)$ solution $= \frac{25.0}{1000} = 0.0250\ dm^3$

moles of $Na_2CO_3 = 0.0250 \times 0.164 = 4.10 \times 10^{-3}$ ✓

moles of HCl $= 4.10 \times 10^{-3}$ ✗

concentration $= 4.10 \times 10^{-3} \times 0.0287 = 1.18 \times 10^{-4}$ ✗

(ii) molar mass of HCl $= 1 + 35.5 = 36.5\ g\ mol^{-1}$

concentration in g $dm^{-3} = 36.5 \times 1.18 \times 10^{-4}$

$= 4.29 \times 10^{-3}\ g\ dm^{-3}$ ✓ [4 marks]

[Total 10 marks]

How to score full marks

(a) (i) The correct equation is:

$$Fe_2O_3 + 3CO \rightarrow 2Fe + 3CO_2$$

Equations must have the same number of atoms of an element on each side of the equation. The student's answer had 2 Fe atoms on the left but only one on the right. **Always check that your equations balance**.

(ii) The student laid out the calculation very clearly, showing all the steps, but then lost a mark **for giving the answer to far too many significant figures**. The answer should be given to the **same number of significant figures as the data**. Here that number is three, and so the correct answer is $1.01 \times 10^3\ dm^3$ (not $1010\ dm^3$)

If you have difficulty with working out the correct number of significant figures, give your answer to **three figures**.

(iii) The correct answer is $11.25 \times 6.02 \times 10^{23} \times \mathbf{2} = 1.35 \times 10^{24}$ atoms.
The student failed to multiply by 2, the number of Fe atoms in Fe_2O_3.

(b) The correct equation is

$$Fe_2O_3 + 3H_2SO_4 \rightarrow \mathbf{Fe_2(SO_4)_3} + 3H_2O$$

The compound is iron (**III**)sulphate. Here the iron has a charge of 3+ and the SO_4 of 2–, so the formula is $Fe_2(SO_4)_3$, which is neutral (6+ and 6–).

2 (a) (i) The correct answer is $\frac{68}{1000} \div 24 = 0.0028$ mol.

The molar volume (24 dm^3 at room temperature and pressure) is given in dm^3, **so you must first convert the volume of 68 cm^3 to dm^3 by dividing by 1000.**

(b) The last line of the answer contains a mistake. The relative molecular mass of the crystals, which have the formula $Na_2CO_3.\mathbf{10H_2O}$, is $[2 \times 23 + 12 + 3 \times 16 + 10 \times (2 + 16)] = 286$, so the mass needed is $0.0410 \times 286 = 11.7$ g.

(c) (i) The first mark is awarded for calculating the moles of sodium carbonate correctly using the concentration and volume.

The second line is wrong. The equation has a ratio of 2HCl for each Na_2CO_3, so moles of HCl = $\frac{\mathbf{2}}{\mathbf{1}}$ × moles of $Na_2CO_3 = \frac{2}{1} \times 4.10 \times 10^{-3} = 8.20 \times 10^{-3}$

The third line is also wrong. **Concentration is moles/volume** and so the 8.20×10^{-3} must be **divided** by 0.0287.

∴ concentration of HCl solution = $\frac{8.20 \times 10^{-3}}{0.0287} = 0.286$ mol dm^{-3}.

(ii) **The mark was awarded here for the correct chemical process.** To go from concentration in moles dm^{-3} to g dm^{-3}, you must multiply by the relative molecular mass, and the student here did this correctly.

Don't make these mistakes ...

Don't forget to check that your **equation** is correct (see the Key Points section on pages 19–20).

Don't round up numbers in the middle of a calculation and don't give your final answer to the wrong number of significant figures.

The **molar volume** of a gas is sometimes given in dm^3 mol^{-1} and sometimes in cm^3 mol^{-1}. Your gas volume must be in the same units.

Don't forget the **stoichiometry** (the numbers of each substance in the equation). To convert moles of chemical A to moles of chemical B, the following has to be calculated:

moles of B = moles of A x the ratio B/A in the equation.

When a **concentration** is in mol dm^{-3} (sometimes called molarity, symbol M), **any volume must be in dm^3**, so convert a volume given in cm^3 to dm^3 by dividing by 1000.

Don't confuse **volume units.**

Don't write down a series of numbers in a calculation. **Set your work out** showing what chemical each number refers to, for example:

amount of $Fe_2O_3 = \frac{\text{mass}}{M_r} = \frac{1.75}{160}$

= 0.01094 mol

or % Fe = 37.1%

Key points to remember

Amount of substance

Moles (n) = $\frac{\text{mass (m)}}{\text{relative molecular mass}}$.

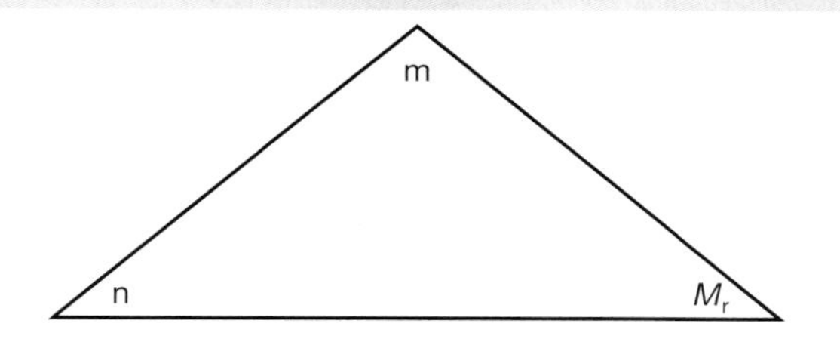

What is the number of moles of Na in 1.6 g of Na?

$\frac{\text{mass}}{M_r} = \frac{1.6}{23} = 0.070$

How many moles of Na in 1.6g of NaOH?

$\frac{\text{mass}}{M_r} = \frac{1.6}{40} = 0.040$

Moles (n) of a gas = $\frac{\text{volume of gas}}{\text{molar volume}}$

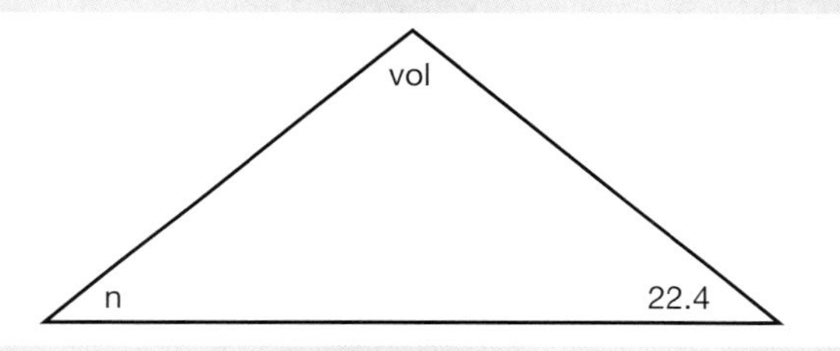

The molar volume of a gas is 22.4 dm^3 at standard temperature and pressure or 24 dm^3 at room temperature and pressure.
How many moles of $O_2(g)$ in 1.6 dm^3 at STP?

volume/molar volume = $\frac{1.6}{22.4} = 0.071$

How many moles of $O_2(g)$ in 1.6 dm^3 at RTP?

volume/molar volume = $\frac{1.6}{24} = 0.067$

Moles (n) of solute in a solution =

concentration in mol dm^{-3} × $\frac{\text{volume in cm}^3}{1000}$

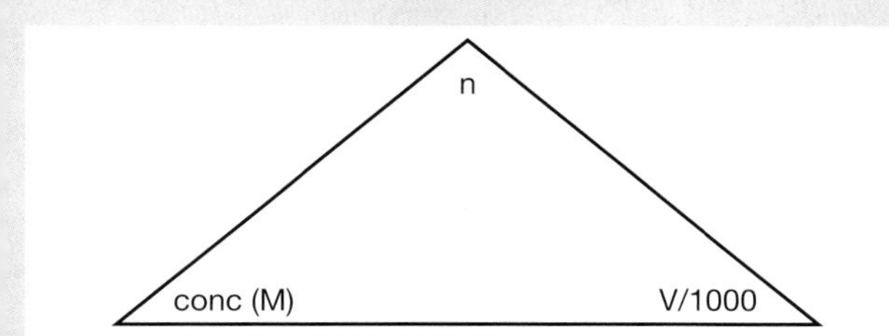

How many moles of NaOH in 23.4 cm^3 of 0.123 mol dm^{-3} solution?

concentration × V(in dm^3) = $0.123 \times \frac{23.4}{1000}$
= 2.88×10^{-3}

What is the concentration of the solution produced when 0.22 mol of NaOH is dissolved in 0.11 dm^3 of water?

moles/volume(in dm^3)= $\frac{0.22}{0.11}$ = 2.0 mol dm^{-3}

Formulae and equations

Empirical formulae

These are the simplest whole number ratio of atoms in the compound. They can be calculated from % composition.

Give values in column 2 to 2 decimal places.

Give values in column 3 to 1 decimal place.

If values in column 3 end in 0.5 multiply all by 2.

Calculate in three columns:

Element and %	% ÷ A_r	÷ smallest
C 48.7%	÷ 12 = 4.06	÷ 2.7 = 1.5
H 8.1%	÷ 1 = 8.10	÷ 2.7 = 3.0
O 43.2%	÷ 16 = 2.70	÷ 2.7 = 1.0

Multiply all by 2 because one ends in 0.5

The empirical formula is $C_3H_6O_2$

Key points to remember

Calculations

There are three main types of calculation:

mass of A → mass of B
mass of A → volume of gas B (or vice versa)
titration of A with B

For all three follow the same route

1. Write a balanced equation.
2. Calculate the moles of A.
3. Moles of B = moles of A × the ratio $\frac{B}{A}$.
4. Calculate the answer.

For steps 2 and 4 use the formula shown above under the heading amount of substance.

For step 3 the ratio $\frac{B}{A}$ is the ratio of the numbers of molecules of B/number of molecules of A in the balanced equation that you wrote in step 1.

Now let's see some examples:

(a) Calculate the mass of H_2SO_4 that will react with 1.25 g of Al.

1 $2Al + 3H_2SO_4 \rightarrow Al_2(SO_4)_3 + 3H_2$

2 moles of Al = $\frac{1.25}{27} = 0.0463$

3 moles of $H_2SO_4 = 0.0463 \times \frac{3}{2} = 0.0694$

4 mass of $H_2SO_4 = 98 \times 0.0694 = 6.81$ g

(b) Calculate the volume of CO_2(at STP) that will react with 5.21 g of NaOH.

1 $2NaOH + CO_2 \rightarrow Na_2CO_3 + H_2O$

2 moles of NaOH = $\frac{5.21}{40} = 0.130$

3 moles of $CO_2 = 0.130 \times \frac{1}{2} = 0.0651$

4 volume of $CO_2 = 22.4 \times 0.0651 = 1.46\ dm^3$

(c) Calculate the volume of 0.333 mol dm^{-3} NaOH that will react with 22.2 cm^3 of 0.0555 mol dm^{-3} H_2SO_4 solution

1 $2NaOH + H_2SO_4 \rightarrow Na_2SO_4 + 2H_2O$

2 moles of $H_2SO_4 = 0.0555 \times \frac{22.2}{1000} = 1.23 \times 10^{-3}$

3 moles of NaOH = $1.23 \times 10^{-3} \times \frac{2}{1}$

4 volume of NaOH = $\frac{2.46 \times 10^{-3}}{0.333} = 7.39 \times 10^{-3}\ dm^3$

$= 7.39\ cm^3$

Molecular equations

Check that all the formulae are correct.

Check that there are the same number of atoms of each element on each side.

If not change the numbers in front of the formulae until the equation balances.

Do not alter any formulae.

Consider this reaction as an example:

Calcium nitrate decomposes into calcium oxide, nitrogen(IV) oxide and oxygen.

Try: $Ca(NO_3)_2 \rightarrow CaO + 2NO_2 + O_2$

Here is the check list:

Formulae: all correct

Numbers of atoms on each side:

Ca 1 and 1, N 2 and 2, O 6 and 7.

The equation does not balance.

Now try $2Ca(NO_3)_2 \rightarrow 2CaO + 4NO_2 + O_2$

Check again:

Ca 2 and 2, N 4 and 4, O 12 and 12

The equation now balances.

Molecular formulae

These are the exact numbers of each atom in the molecule.

If an empirical formula is CH_2 and the relative molecular mass is 56, the molecular formula is C_4H_8, because CH_2 has a mass of 14 and $14 \times 4 = 56$.

Ionic equations

Apart from following the rules for molecular equations, you must also check that the total charge on the left equals the total charge on the right.

Chlorine oxidises Fe^{2+} ions to Fe^{3+} ions:

Try $Cl_2 + Fe^{2+} \rightarrow 2Cl^- + Fe^{3+}$

The formulae are correct, numbers of atoms balance, but the charge on left is 2+ while the charge on right is 2– + 3+ = 1+.

The correct equation is:

$$Cl_2 + 2Fe^{2+} \rightarrow 2Cl^- + 2Fe^{3+}$$

Now charge: left = 4+, right is 2– + 6+ = 4+

Questions to try

Examiner's hint ***for question 1***

(a) moles = $\frac{\text{mass}}{M_r}$ or for a gas, moles = $\frac{\text{volume}}{\text{molar volume}}$

moles of solute in a solution = $\frac{\text{volume (V) in cm}^3}{1000}$ × concentration (M or molarity) in mol dm^{-3}

Calcium and hydrochloric acid react according to the equation:

$$Ca(s) + 2HCl(aq) \rightarrow CaCl_2(aq) + H_2(g)$$

(a) 0.122 g of calcium metal was added to 25.0 cm^3 of 1.23 mol dm^{-3} solution of hydrochloric acid and the hydrogen gas was collected at room temperature and pressure.

(i) Calculate whether the calcium or the hydrochloric acid was in excess.

....

....

....

.... [4 marks]

(ii) Hence calculate the volume of hydrogen gas produced. (1 mol of gas under these conditions occupies 24 000 cm^3.)

....

....

[2 marks]

(b) Calcium metal often contains an impurity of calcium oxide on its surface. This also reacts with hydrochloric acid, but does not produce a gas. In the experiment in (a), it was found that only 70.1 cm^3 of hydrogen was produced. Calculate the percentage purity of the calcium metal.

....

....

[2 marks]

(c) Hard water contains calcium ions. The amount can be found by adding an excess of sodium ethandioate and weighing the precipitate formed. Sodium ethandioate contains the following percentages by mass:

sodium 34.3%, carbon 17.9% oxygen 47.8%

(i) Calculate the empirical formula of sodium ethandioate.

....

....

....

[3 marks]

(ii) The relative molecular mass of sodium ethandioate is 134. Use this to deduce its molecular formula.

....

....

[2 marks]

[Total 12 marks]

Examiner's hints ***for question 2***

(a) Moles of gas B = moles of gas A × ratio B/A in the equation.

(b) 1 mole contains 6.02×10^{23} N molecules or ion groups.

So 1 mole of NH_3 contains 6.02×10^{23} N atoms and $3 \times 6.02 \times 10^{23}$ H atoms.

Q2

Propane, C_3H_8, is used as a fuel when liquefied in containers under pressure. When the tap on the container is opened, the propane comes out as a gas.

(a) (i) Write a balanced equation for the complete combustion of propane gas into carbon dioxide and water.

....

[2 marks]

(ii) Calculate the volume of oxygen and hence of air that is required to burn 25 cm^3 of propane gas at room temperature and pressure. (Air contains 20% oxygen).

....

.... [3 marks]

(b) Calculate

(i) the number of moles of carbon dioxide in 75 cm^3 of carbon dioxide gas measured at room temperature and pressure. (1 mole of gas occupies 24 dm^3 under these conditions.)

....

[1 mark]

(ii) the number of oxygen atoms in this amount of carbon dioxide. (The Avogadro constant is 6.02×10^{23} mol^{-1})

....

.... [2 marks]

(iii) the concentration, in mol dm^{-3}, of dissolved carbon dioxide if this volume of gas dissolves in 1.5 dm^3 of solution.

....

[1 mark]

(c) Carbon dioxide reacts with aqueous sodium carbonate to produce sodium hydrogen carbonate. The equation is:

$$Na_2CO_3 + H_2O + CO_2 \rightarrow 2\ NaHCO_3$$

(i) Calculate the mass of sodium hydrogen carbonate formed when 4.56 g of sodium carbonate reacts.

....

.... [3 marks]

(ii) Sodium hydrogen carbonate is a neutral substance that reacts with acids. If 100 cm^3 of 5.00 mol dm^{-3} sulphuric acid was spilt, calculate the minimum mass of sodium hydrogen carbonate that would be needed to neutralise it.

The equation is

$H_2SO_4 + 2NaHCO_3 \rightarrow Na_2SO_4 + 2H_2O + 2CO_2$

....

....

.... [3 marks]

[Total 15 marks]

Examiner's hint ***for question 3***

When balancing an equation:

1. Check there are the same numbers of each element on each side of the equation.
2. For ionic equations, check that the total charge on the left-hand side of the equation equals the total charge on the right-hand side.

If either is wrong, alter the numbers in front of the formulae until the equation balances.

Do not alter formulae.

Q3

Balance the following equations

(a) $Al_2O_3 + HCl \rightarrow AlCl_3 + H_2O$ [1 mark]

(b) $LiNO_3 \rightarrow Li_2O + NO_2 + O_2$ [1 mark]

(c) $K + H_2O \rightarrow KOH + H_2$ [1 mark]

(d) $Ca(OH)_2 + Na_3PO_4 \rightarrow Ca_3(PO_4)_2 + NaOH$ [2 marks]

(e) $Fe^{3+}(aq) + Sn^{2+}(aq) \rightarrow Fe^{2+}(aq) + Sn^{4+}(aq)$ [1 mark]

(f) $I^-(aq) + IO_3^-(aq) + H^+(aq) \rightarrow I_2(aq) + H_2O(l)$

[2 marks]

The answers to these questions are on pages 78–80.

[Total 8 marks]

Exam Questions and Student's Answers

1 Sodium metal, sodium fluoride and hydrogen fluoride all form different types of solids with different bonding.

(a) State the type of solid in sodium, sodium fluoride and hydrogen fluoride.

Sodium is metallic, sodium fluoride ionic and hydrogen fluoride is molecular.

[3 marks] 

(b) Describe the bonding in metallic sodium and hence explain why it conducts electricity in the solid state.

Each sodium atom loses its s electron which form a sea of delocalised electrons which are free to move and so the solid conducts electricity.

[4 marks] 

(c) Describe the bonding in sodium fluoride and explain why it conducts electricity when molten but not when solid.

A sodium atom gives an electron to a fluorine atom so that both reach the stable electronic structure of a noble gas.
In the liquid the ions are free to move around.

[4 marks] 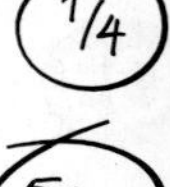

2 (a) Explain, in terms of the intermolecular forces involved, why ethanol, C_2H_5OH, has a much higher boiling point than methoxymethane, CH_3OCH_3.

There are hydrogen bonds in ethanol and these are stronger than the van der Waals forces between methoxymethane molecules.

[4 marks] 

(b) (i) State the bonding in lithium fluoride.

Ionic ✓

[1 mark] 

(ii) Explain, in terms of the arrangement and movement of particles, what happens when solid lithium fluoride is heated from room temperature to just above its melting temperature.

When it is heated the atoms ✗ begin ✗ to vibrate. Eventually the vibrations are so large that the lattice ✓ breaks up ∧ and the substance becomes a liquid.

[3 marks] 

[Total 8 marks]

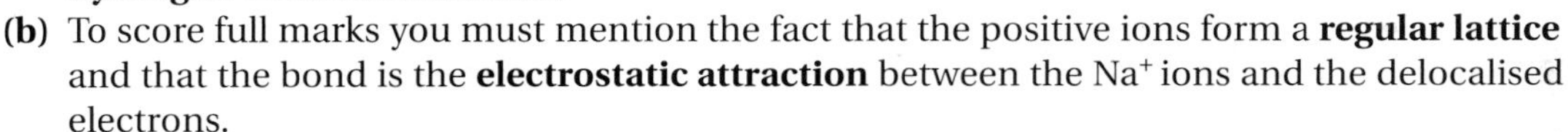

How to score full marks

(a) The correct answer is that hydrogen fluoride forms a **hydrogen bonded molecular solid.** A mark is lost here because the phrase 'hydrogen fluoride is molecular' is ambiguous. **There are two types of molecular solid; simple and hydrogen bonded molecular**.

(b) To score full marks you must mention the fact that the positive ions form a **regular lattice** and that the bond is the **electrostatic attraction** between the Na^+ ions and the delocalised electrons.

(c) At this level, it is correct to describe an ionic bond as **the force of attraction between oppositely charged Na^+ and F^- ions** that results from the **loss** of an electron from the sodium atom and the **gain** of an electron by the fluorine atom.

The question is in two parts and the student ignored some of the second part, and did not explain why the solid does not conduct. The correct answer is that the ions in **solid** sodium fluoride are fixed in position and are **not free to move**, and so **cannot conduct** an electric current.

(a) In questions on melting or boiling temperatures you need to make two points:

- Identify the **types** of forces between the particles for both substances.
- Then, state the **relative strengths** of these forces.

You must make it clear that hydrogen bonds are **between** molecules. In this question they are between the δ^+ **H** of the OH group in one ethanol molecule and the δ^- **O** in another.

A correct answer is:

In ethanol there are hydrogen bonds between the molecules (see diagram) and these are stronger than the intermolecular van der Waals forces between the molecules of methoxymethane.

$$\begin{array}{ccccc} & & & & H \\ & & & & / \\ O & — & H^{\delta+} & \cdots\cdots & O^{\delta-} \\ / & & & & \backslash \\ C_2H_5 & & & & C_2H_5 \end{array}$$

Note that it is a good idea to draw a diagram as this shows the hydrogen bond clearly. **Both** the permanent dipole/dipole and the instantaneous induced dipole/induced dipole (also called dispersion) forces are normally called **van der Waals** forces.

(b) (ii) There are two errors in the first sentence:

- The particles are **ions** not atoms.
- The ions do **not start** to vibrate. They are already vibrating at any temperature.

The correct thing to say is that the vibration of the ions **increases** as the temperature rises. The second mark was gained for stating that the **lattice breaks up**. To gain the third mark you must explain that **the ions move around** freely after the solid has melted.

Don't make these mistakes ...

In your explanation of an ionic bond do **not** write about noble gas structures. An ionic bond is the **attraction** between **oppositely** charged ions formed by the transfer of electrons.

Do not state that covalent bonds are weak. They are about the **same** strength as ionic bonds. The weak forces/bonds are intermolecular forces – those **between** molecules.

Do not confuse intermolecular forces (hydrogen bonds and van der Waals forces) with covalent bonding. The former are **between** molecules and the latter are **within** molecules.

Do not forget to count **lone pairs** when deducing shapes of molecules or ions. The number of lone pairs can be calculated using the formula:

Number of lone pairs

$$= \frac{\text{(group number of the atom – the number of covalent bonds)}}{2}$$

But for a negative ion add 1 electron per negative charge to the group number.

For a positive ion subtract 1 electron per positive charge from the group number.

When deducing bond angles, do not forget **lone pairs**. Thus, ammonia, which has **one lone pair**, is pyramidal with a bond angle of less than 109.5° and water, which has **two lone pairs**, is V shaped with a bond angle less than that of ammonia.

Key points to remember

Chemical bonds

An **ionic bond** is the electrostatic attraction between a positive ion (the cation) and a negative ion (the anion). The strength depends on two factors:

- The charge on the ions – the larger the stronger the bond
- The sum of the ionic radii – the smaller the stronger the bond.

Cations are formed by the **loss** of one or more electrons. e.g. Na (2,8,1) forms Na^+ (2,8).

Anions are formed by the **gain** of one or more electrons. e.g. Cl (2,8,7) forms Cl^- (2.8.8).

Polarisation

The electrons on an **anion** can become polarised and this causes the bond to have some covalent character.

LiI is partially covalent, whereas LiF is totally ionic. A cation that is **small** and/or is **highly charged** has a large polarising power. e.g. the doubly charged Mg^{2+} ion is more polarising than the singly charged Na^+ ion.

An anion that is **large** and/or is **highly charged** is easily polarised. e.g. the large I^- is more easily polarised than the smaller F^- ion.

A **covalent** bond is the **sharing of a pair** of electrons by two atoms.

It is formed when atomic orbitals, each containing **one** electron, **overlap**.

A **short** covalent bond is **stronger** than a longer bond, e.g. the C—Cl bond is stronger than the C—I bond.

A π bond is weaker than a σ bond.

The C=C bond strength is less than twice the C—C bond strength.

A σ bond is formed by the 'head on' overlap of 2 orbitals, usually two p orbitals. The result is an electron cloud that is on a line between the nuclei:

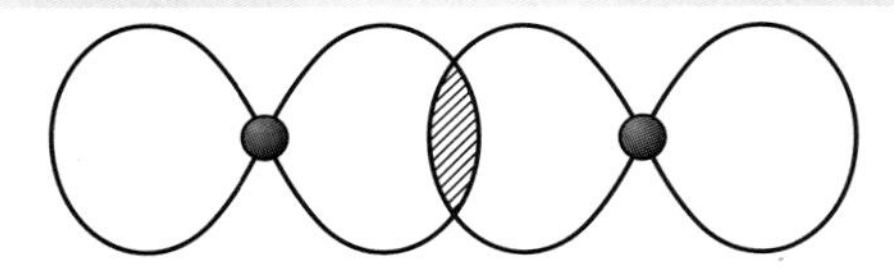

A π bond is formed by the 'sideways' overlap of two p orbitals, resulting in the electron cloud being above and below the line between the nuclei:

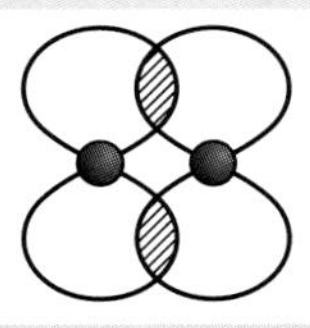

A **dative covalent (co-ordinate)** bond is formed when one of the atoms provides **both** electrons to the shared pair.

The result is a shared pair of electrons as in an ordinary covalent bond.

The ammonium ion contains one dative and three covalent bonds.

H
××
H ×o N ×o H
×o
H
+

A **polar** covalent bond results if the electronegativities of the two atoms are different.

The electronegativity of an element is defined as the extent to which it draws the electrons in a covalent bond towards itself.

The electronegativities increase across a period and decrease down a group. Thus fluorine is the most electronegative element.

HCl and HBr are both polar molecules, but in totally symmetrical molecules, such as CH_4 and CO_2, the polarity of each bond cancels out and the molecule is itself not polar.

Key points to remember

A **metallic** bond occurs when the metal atoms lose their outer electrons. These then form a 'sea' of delocalised electrons in a regular lattice of positive ions.

The bond is the force of attraction between the delocalised electrons and the positive ions.

The electrons in a metal are able to move through the lattice and so the solid conducts electricity.

Intermolecular forces

The bonds **between** covalent molecules are called **intermolecular** forces and are much weaker than the strong covalent bonds **within** the molecule.

Types of solid	Examples
Ionic	NaCl, MgO, $Ba(OH)_2$
Metallic	All of groups 1 and 2 and aluminium, tin and lead and the *d*-block elements
Molecular: simple	I_2, CO_2
hydrogen bonded	HF, H_2O, NH_3, alcohols
Giant atomic (or giant covalent or giant molecular)	Graphite, diamond, silicon and quartz (SiO_2)

Hydrogen bonds

A hydrogen bond occurs between a δ+ H atom in one molecule and a δ− F, O or N atom in another molecule.The hydrogen bonds in ice are the dotted lines and the covalent bonds are the solid lines:

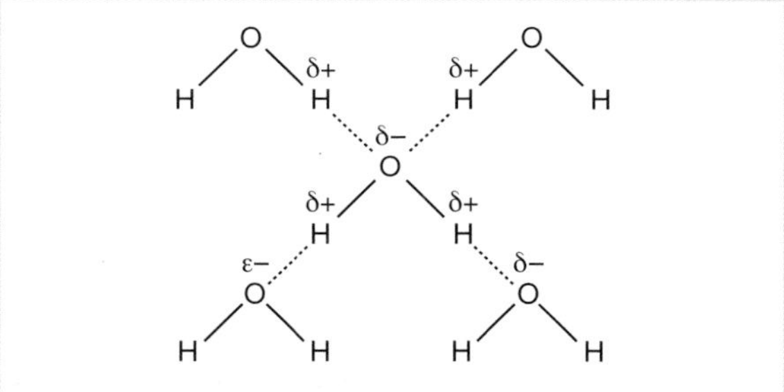

H O δ+ H δ+ H O H
δ− O
δ+ H δ+ H
ε− O δ− O
H H H H

Change of state

Heating a solid until it just liquefies.

The **stronger** the intermolecular force the **higher** the melting point.

As a solid is heated the particles **vibrate more** until the energy is sufficient to **break up** the lattice, the solid melts and the particles become **free to move** about.

Instantaneous induced dipole/induced dipole (or dispersion force)

This is significant for larger molecules. The strength depends mainly on the total **number** of **electrons** in the molecule. This dispersion force in HBr is stronger than the dispersion force in HCl because HBr has 36 electrons and HCl has only 18.

HCl has a higher boiling point than Ar even though the dispersion forces are about the same strength, because HCl is also polar.

Permanent dipole/dipole force

This is significant if the molecule is small and very polar.

A good example is HCl, which is very polar:

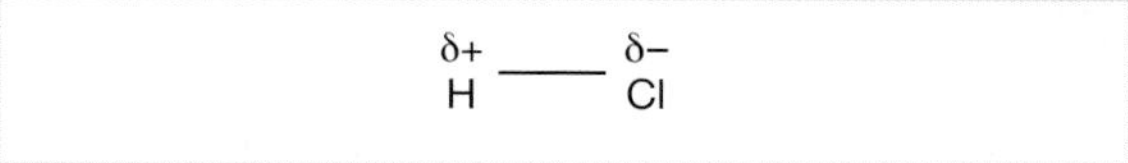

$\overset{\delta+}{H} — \overset{\delta-}{Cl}$

Heating a liquid until it boils

The **stronger** the intermolecular force the **higher** the boiling point.

As a liquid is heated, its particles gain kinetic energy and are more likely to break free from the forces holding them in the liquid. They therefore escape from the surface by evaporation.

As the boiling point is reached, the particles gain sufficient energy to escape from the inside of the liquid.

Shapes of molecules and ions

The shape of a molecule or ion is determined by the number of **bond** pairs and **lone** pairs of electrons around the central atom. The electron pairs repel each other and get as far apart as possible. The lone pair/ lone pair repulsion is greater than lone pair / bond pair, which is greater than bond pair / bond pair.

l.p/l.p > l.p/b.p > b.p/b.p.

The bond angles depend on the arrangement of electrons and the number of lone pairs. In a tetrahedral molecule it is 109.5°. However, a lone pair reduces that angle, and two lone pairs reduce it further. The following table shows some bond angles:

Numbers of pairs	Shape	Example	Bond angle
2 bond pairs	linear	$BeCl_2$  Cl — Be — Cl	180°
3 bond pairs	triangular	$AlCl_3$ 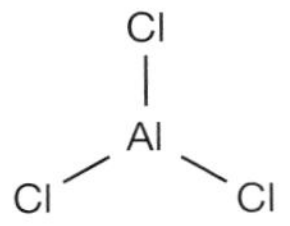	120°
4 bond pairs	tetrahedral	CH_4 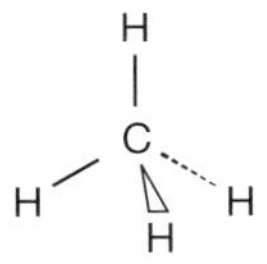	109.5°
3 bond + 1 lone pair	pyramidal	NH_3 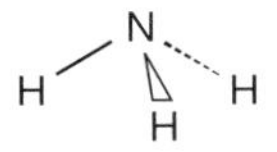	107°
2 bond + 2 lone pairs	bent (non-linear)	H_2O 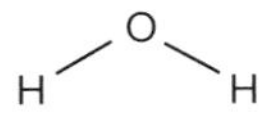	104.5°
5 bond pairs	triangular bipyramid	PCl_5 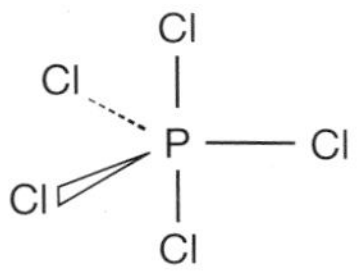	120° and 90°
6 bond pairs	octahedral	SF_6 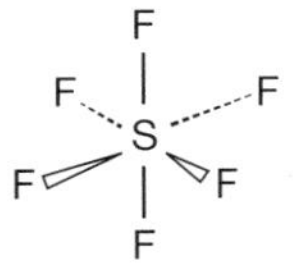	90° and 180°

Questions to try

Examiner's hints ***for question 1***

(a) In a 1+ ion, an atom has lost 1 electron.
In a 1− ion, an atom has gained 1 electron.
The order of repulsion is: l.pr/l.pr > l.pr/b.pr > b.pr/b.pr.

(b) Don't forget any lone pairs.

Q1

(a) Explain the principles that determine the shapes of molecules.

....

....

....

[3 marks]

(b) Draw dot and cross diagrams (showing outer electrons only) and explain the shapes of the following:

(i) NH_4^+

[3 marks]

(ii) NH_3

[3 marks]

(iii) NH_2^-

[3 marks]

(iv) Hence predict which has the greatest HNH bond angle, NH_4^+, NH_3 or NH_2^-.

....

....

[2 marks]

(c) A double bond is regarded as one region of negative charge and so the number of π bonds is ignored when determining the shape. Draw the shapes, name them and state the bond angle in:

(i) CO_2 **(ii)** SO_2

[6 marks]

[Total 20 marks]

Examiner's hints *for question 2*

(a) Hydrogen bonds are drawn as dotted lines and covalent bonds as solid lines.

(b) First identify the type of intermolecular forces and then state their relative strengths. Don't forget to compare HF with HCl and then HCl with HBr.

Q2

(a) What is meant by the term hydrogen bond? Illustrate your answer by reference to the bonding in water.

....

....

[4 marks]

(b) Explain why HF has a higher boiling temperature than HCl, but HCl has a lower boiling temperature than HBr.

....

....

.... [4 marks]

[Total 8 marks]

Examiner's hints *for question 3*

(a) The polarities of the bonds may cancel out in the molecule.

(b) You must make it clear which factor is the vital one in each case.

Q3

(a) Explain why HCl is a polar molecule whereas Cl_2 and CCl_4 are not.

....

....

.... [5 marks]

(b) State and explain which of each pair is the more polarising:

(i) Li^+ or Na^+ ....

.... [2 marks]

(ii) Na^+ or Mg^{2+} ....

.... [2 marks]

(c) Explain why AlF_3 is ionic whereas $AlCl_3$ is covalent.

....

.... [3 marks]

The answers to these questions are on pages 80–82. [Total 12 marks]

Exam Question and Student's Answer

1 (a) The periodic table is divided into groups and periods. What is the difference in electronic configuration between:

(i) two elements one above the other in a group?

They have the same number of electrons in their outer shells ✗

[1 mark] 0/1

(ii) two adjacent elements in a period?

One has one more electron than the other. ✓

[1 mark] 1/1

(b) Explain, in terms of structure and bonding, why lithium and graphite conduct electricity in the solid state.

Lithium is a metal. Its outer 2s electron is part of a delocalised ✓ cloud of electrons between a regular arrangement of positive Li^+ ions. These electrons can move through the solid. ✓

In graphite, the π electrons are in a delocalised ✓ cloud above and below the plane ✓ of the molecule. They also can move and so it conducts electricity.

[5 marks] 4/5

(c) The melting points of the elements in period 3 are shown below:

	sodium	magnesium	aluminium	silicon	phosphorus	sulphur	chlorine	argon
m.pt/°C	98	650	660	1410	44	119	– 101	– 189

Explain why the melting point of:

(i) aluminium is greater than that of sodium.

Aluminium has three delocalised electrons, and so the forces between the delocalised electrons and the Al^{3+} ion is greater than between the delocalised electrons and the Na^{+} ion.

[3 marks] 

(ii) silicon is the highest in the group.

The melting point reaches a maximum at silicon because it can form 4 bonds. Sodium and chlorine form one, magnesium and sulphur two, aluminium and phosphorus three, so the bonding in silicon is the strongest.

[4 marks] 

(iii) sulphur is higher than that of phosphorus.

The bonding in sulphur is stronger because, being smaller, it forms stronger covalent bonds.

[3 marks] 

(d) Some data is shown below about silicon:

electronegativity	1.8
atomic radius / nm	0.117
1st ionisation energy / kJ mol^{-1}	786

State and explain which of aluminium or phosphorus has:

(i) an electronegativity of 2.1

Phosphorus, because the electronegativity increases left to right in the periodic table as the atoms become smaller, and so the nucleus has a greater pull on the bonding electrons. ✓✓

[2 marks] 2/2

(ii) an atomic radius of 0.110 nm.

Aluminium, as the atoms get bigger as they have more electrons. ✗✗

[2 marks] 0/2

(iii) a 1st ionisation energy of 577 kJ mol^{-1}.

Aluminium, ✓ as the 1st ionisation energies increase across a period as the nuclear charge increases without ✓ an increase in the number of inner (shielding) electrons.

[2 marks] 2/2

[Total 23 marks]

11/23

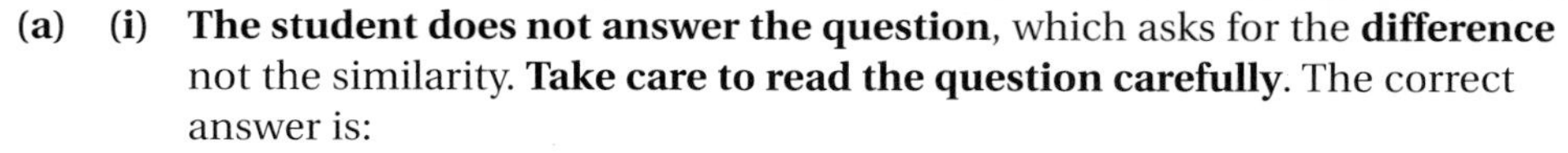

How to score full marks

(a) (i) The student does not answer the question, which asks for the **difference** not the similarity. **Take care to read the question carefully**. The correct answer is:

The element higher in the group has **one fewer electron shell** than the one lower in the group.

(b) The explanation of why lithium conducts electricity is correct, but the answer for graphite is incomplete. The correct answer is:

The structure of graphite consists of **interlocking hexagons** in a plane. One electron from each carbon atom forms a **delocalised** (π) **cloud above and below** the plane. **These delocalised electrons can also move through the solid and so graphite conducts electricity**.

(c) (i) The student omitted one important reason why aluminium has the higher melting point. He or she should have also written:

The Al^{3+} ion is much smaller than the Na^{+} ion. This means that the metallic bond is much stronger and so the melting point is higher.

(ii) The answer here is totally wrong. The correct answer is:

Silicon forms a **giant atomic (covalent or molecular)** lattice. All the **covalent bonds** must be broken before silicon will melt. Covalent bonds are much **stronger than intermolecular** forces in the other non-metals, and **stronger than the metallic** bond in the s and p block metals, so silicon has the highest melting point.

(iii) This answer is also totally wrong. The correct answer is:

Sulphur forms $\mathbf{S_8}$ molecules and phosphorus forms $\mathbf{P_4}$ molecules. There **are more electrons** in a sulphur molecule than in a phosphorus molecule, so the **intermolecular** forces are stronger, hence sulphur has a higher melting point.

Note: when a molecular solid melts, it is the **inter**molecular forces that are broken, **not** the covalent bonds between atoms **within** the molecule

(d) (ii) The correct answer is phosphorus, as the atoms get **smaller** across a period because the nuclear charge **increases**.

Don't make these mistakes ...

Don't think that the atoms get larger across a period. They get **smaller** because, as the nuclear charge increases, the electrons get pulled in more.

Don't get the formulae of **group II compounds** wrong:
Oxides are MO
Hydroxides are $M(OH)_2$
Carbonates are MCO_3
Sulphates are MSO_4
Nitrates are $M(NO_3)_2$

Don't forget that the solubility of **sulphates decreases** down group II and the solubility of **hydroxides increases** down group II.

Don't confuse **intermolecular** forces, which are **weak**, with **covalent** bonds, which are **strong**.

Don't forget to learn the formulae of the **non-metallic elements** in periods 2 and 3.

Don't forget to check that the formulae are correct when writing an equation. Check also that the **equation balances**.

Key points to remember

The Periodic Table

- Elements are arranged according to **atomic number**.
- The periodic table is divided into horizontal rows called **periods**.
- It is divided into vertical columns called **groups**.
- The **s** block consists of groups I and II.
- The **p** block consists of groups III, IV, V, VI, VII and the noble gases (group 0).
- The **d** block is the elements scandium to zinc and those below them.

Trends across a period

Atomic radius	This **decreases** left to right because the nuclear charge increases without an increase in inner (shielding) electrons.
Electronegativity	This **increases** left to right because the smaller atoms exert a greater pull on the bonding electrons.
1st ionisation energy	The general trend is to **increase** left to right as the nuclear charge increases without any increase in shielding.
Electrical conductivity	**Metals** (and **graphite**) conduct; non-metals do not conduct.

Melting and boiling points

To explain the variation in melting and boiling points, you must:

- Decide on the **type of solid** (metallic, giant atomic/covalent/molecular or simple molecular).
- Decide on the **type of force** (due to delocalised electrons, covalent bond or van der Waals force).
- Comment on the **strength of the force** (covalent bond > metallic bond in s and p block > van der Waals force).

Strength of force depends on various factors:

- If metallic, the strength depends on the **number of delocalised electrons and the size** of the metal ion (the smaller the ion, the higher the melting point).
- If giant atomic/covalent/molecular, the strength depends on the **length of the covalent bonds**. Shorter bonds are stronger; the C–C bond length is less than the Si-Si and so the C–C bond is stronger.
- If molecular, the strength depends on the **number of electrons in the molecule**. Therefore the melting (and boiling) points are in the order:

$S_8 > P_4 > Cl_2 > Ar$

$F_2 > O_2 > N_2 > Ne$

Trends in Group II (Group 2) of the Periodic Table

Atomic radius	**Increases** down a group as the number of shells of electrons increases.
Electronegativity	**Decreases** down a group as the atoms become larger.
1st ionisation energy	**Decreases** down a group as the outer electron becomes further away from the nucleus. More shielding occurs with increase in nuclear charge.
Reactivity	Reactivity as a metal **increases** down the group because it becomes easier to remove two electrons from the outer shell.

Reactions of Group II elements and compounds

Reaction of element (M) with oxygen	All burn to form an oxide: $2M + O_2 \rightarrow 2MO$
Reaction of element with water	Be does not react. Mg burns in steam: $Mg + H_2O \rightarrow MgO + H_2$ The others react rapidly with cold water: $Ca + 2H_2O \rightarrow Ca(OH)_2 + H_2$
Reaction of oxide (MO) with dilute hydrochloric acid	All react as bases: $MO + 2HCl \rightarrow MCl_2 + H_2O$
Reaction of carbonate (MCO_3) with dilute hydrochloric acid	All react to form carbon dioxide gas: $MCO_3 + 2HCl \rightarrow MCl_2 + CO_2 + H_2O$
Effect of heat on carbonate	$BaCO_3$ is stable to heat. The stability **increases** down the group. The others decompose: $CaCO_3 \rightarrow CaO + CO_2$
Solubility of sulphate (MSO_4)	The solubility **decreases** down the group. $BaSO_4$ is insoluble and is the precipitate in the test for sulphates.
Solubility of hydroxide ($M(OH)_2$)	The solubility **increases** down the group. $Mg(OH)_2$ is almost insoluble, $Ca(OH)_2$ slightly soluble and $Ba(OH)_2$ fairly soluble.

Questions to try

Examiner's hints ***for question 1***

- Remember to check all formulae in your equations and check that your equations balance.
- Ionic equations do not include spectator ions.
- Check that you know the trends down a group.
- Check that you understand the trends across a period.

Q1

(a) (i) State the trend in solubility of the group II hydroxides.

....

[1 mark]

(ii) A solution of sodium sulphate was added to a solution of a group II chloride and a white precipitate was obtained. Name the group II chloride and write an **ionic** equation for its reaction with aqueous sodium sulphate.

Name ....

Ionic equation ....

[2 marks]

(b) (i) Write the equation for the decomposition of calcium carbonate.

....

[1 mark]

(ii) Calcium carbonate reacts with acids such as dilute hydrochloric acid to liberate carbon dioxide. Write an equation, with state symbols, for this reaction.

....

[2 marks]

(c) Below is some data about the element calcium.

Electronegativity	1.0
Atomic radius /nm	0.174

State and explain which of magnesium and strontium has:

(i) an atomic radius of 0.191 nm

....

....

[2 marks]

(ii) an electronegativity of 1.2

....

....

[2 marks]

(d) Sodium and magnesium are classified as s block elements and the other elements in period 3 as p block.

(i) Write the electronic configuration, using the s, p, d notation, of

magnesium ....

and phosphorus .... [2 marks]

(ii) Explain why the boiling point of sulphur is much higher than that of chlorine.

....

....

.... [3 marks]

(iii) State the formula of a period 3 oxide that is basic

....

[1 mark]

[Total 16 marks]

Examiner's hints ***for question 2***

(a) The explanation for trends depends on:

- The nuclear charge.
- The number of inner (shielding) electrons.
- The radius of the atom or ion.

(b) Metals form positive ions when they react.

Q2

Use the data below to answer the questions that follow:

	[Be]	[Mg]	[Ca]	[Sr]	[Ba]
1st ionisation energy / kJ mol^{-1}	900	736	590	548	502
2nd Ionisation energy / kJ mol^{-1}	1760	1450	1150	1060	966
Metallic radius / pm	112	160	197	215	217

(a) Explain the trends in 1st ionisation energy

....

....

.... [3 marks]

(b) Magnesium reacts very slowly with hot water, but burns when heated in steam.

Calcium reacts rapidly with cold water.

Barium reacts even more rapidly with cold water.

(i) Write the equation (with state symbols) for calcium's reaction with water.

.... [2 marks]

(ii) Write the equation (with state symbols) magnesium's reaction with steam

.... [1 mark]

(iii) Explain the trend in reactivity of the group II metals with water.

....

.... [2 marks]

(c) Radium is in group II and is below barium. Predict:

(i) the reaction between radium and water.

....

.... [2 marks]

(ii) the solubility of radium sulphate.

.... [1 mark]

(iii) the metallic radius of radium.

.... [1 mark]

The answers to these questions are on pages 82 and 83. [Total 12 marks]

Exam Question and Student's Answer

1 (a) $Cr_2O_7^{2-}$ ions in acid solution will oxidise Sn^{2+} ions. The unbalanced ionic equation for this reaction is:

$$Cr_2O_7^{2-}(aq) + H^+(aq) + Sn^{2+}(aq) \rightarrow Cr^{3+}(aq) + Sn^{4+}(aq) + H_2O(l)$$

(i) Evaluate the oxidation numbers of chromium in the $Cr_2O_7^{2-}$ and the Cr^{3+} ions and hence calculate the change in oxidation number per chromium atom in this reaction.

Chromium goes from +6 in $Cr_2O_7^{2-}$ to +3 ✓ in Cr^{3+} so the change is ^ 3 per chromium atom. [2 marks]

(ii) Evaluate the change in oxidation numbers of a tin ion.

The tin ions go from +2 to +4 and so the change is +2 for each tin ✓ ion. [1 mark]

(iii) Hence write the balanced equation for this reaction.

$Cr_2O_7^{2-}(aq) + 14H^+(aq) + 3Sn^{2+}(aq) \rightarrow 2Cr^{3+}(aq) + 3Sn^{4+}(aq) + 7H_2O(l)$ ✓✓

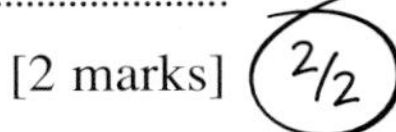

[2 marks]

(b) (i) Write an ionic half equation for Fe^{2+} ions being oxidised to Fe^{3+} ions.

$Fe^{2+} + e^- \rightarrow Fe^{3+}$ [1 mark]

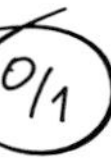

(ii) Write an ionic half equation for chlorine being reduced to Cl^- ions.

$Cl_2 + 2e^- \rightarrow 2Cl^-$ ✓ [1 mark]

(iii) Hence write the overall equation for the reaction between chlorine and Fe^{2+} ions.

$Cl_2 + {}_\wedge Fe^{2+} \rightarrow 2Cl^- + {}_\wedge Fe^{3+}$ [1 mark]

(c) All the halogens are oxidising agents, but chlorine is a stronger oxidising agent than bromine.

(i) Write equations to show that both chlorine and bromine are oxidising agents.

Both will oxidise iodide ions $Cl_2 + 2I^- \rightarrow 2Cl^- + I_2$ ✓

and $Br_2 + 2I^- \rightarrow 2Br^- + I_2$ ✓

[2 marks] 2/2

(ii) Consider the results of adding separate aqueous solutions of chlorine and bromine to solutions containing chloride, bromide and iodide ions. What **observations** would show that chlorine is a stronger oxidising agent than bromine?

Chlorine will oxidise bromide ions to bromine, whereas bromine will not oxidise chloride ions to chlorine. ✗

[2 marks] 

(d) The physical and chemical properties of the halogens alter steadily down the group.

(i) State and explain how the boiling points alter down the group.

The boiling point decreases ✗ down the group, because ✗ the strength of the covalent bond gets less as the atoms get bigger making it easier to break the bonds.

[3 marks] 

(ii) State and explain how the electronegativity changes down the group.

The electronegativity decreases ✓ down the group, as the atomic radius gets larger. This means that the nucleus has less of a pull on the electrons as they are further ✓ from the centre of the halogen atom.

9/18

[3 marks] 2/3

[Total 18 marks]

How to score full marks

(a) **(i)** It is important to indicate the **direction of change in oxidation number**. The student should have given **–3 (or down by 3)** as the answer.

(b) **(i)** If a substance is being **oxidised**, the electrons in a half equation must be on the **right** hand side (OIL RIG – **o**xidation **i**s **l**oss). Thus the corrects answer is:

$$Fe^{2+} \rightarrow Fe^{3+} + e^-$$

(iii) When adding half equations, **the number of electrons must be the same in both.** So equation (i) must be multiplied by 2 and added to equation (ii). The correct overall ionic equation is: $Cl_2 + 2Fe^{2+} \rightarrow 2Cl^- + 2Fe^{3+}$

Ionic equations must also balance for charge. In the student's answer the left-hand side is +2 and the right is only +1. Note that in the correct equation both sides are +4.

(c) **(ii)** The student did not answer the questions that asked for **observations**. The correct answer is:

"Chlorine will turn a solution of colourless bromide ions to a red-brown colour, but the red-brown colour of bromine remains when bromine is added to a solution of chloride ions."

(d) **(i)** This is an example of a common error. The halogens form simple diatomic covalent molecules and it is the **intermolecular forces not the covalent bonds** that have to be broken on boiling. The correct answer is:

"The **intermolecular** forces, which here are instantaneous induced dipole forces*, get **stronger** down the group as the **number of electrons** in the molecule increases. Thus the boiling point of the halogens **increases** down the group."

* These forces are sometimes called London or dispersion forces and are a type of van der Waals force.

(ii) Electronegativity measures the pull on a **pair of bonding** electrons

Don't make these mistakes ...

Don't forget OIL RIG – **O**xidation **I**s **L**oss and **R**eduction **I**s **G**ain

Don't get muddled about redox. Remember that **oxidising agents** (such as chlorine) become **reduced**, and **reducing agents** (such as Fe^{2+} ions and iodide ions) become **oxidised**.

Don't forget that ionic half equations contain electrons. **Make sure that you get the electrons on the correct side**. When a substance is reduced, electrons are gained and so go on the left-hand side of a half equation, and when a substance is oxidised, electrons are lost and so are on the right in a half equation.

Don't forget that the number of electrons in a half equation equals the total change in oxidation number. If the substance is in acid solution you will probably need to have H^+ on the left and H_2O on the right.

Key points to remember

REDOX

Oxidation number or state is the charge that an element would have if all the bonds in the substance were ionic.

- Uncombined elements are zero.
- F is always −1, Group I are +1, Group II +2, oxygen is −2 (except in peroxides), H is +1 (except in metallic hydrides).
- Oxidation numbers in a neutral compound add up to zero and in a polyatomic ion they add up to the charge on that ion.

Some examples:

- Mn in MnO_4^- is +7 as each O is −2 = −8, so Mn + (−8) = −1, therefore Mn is +7.
- Cr in $Cr_2O_7^{2-}$ is +6 as each O is −2 = −14, so 2Cr + (−14) = −2, therefore each Cr is +6.
- Cl in ClO^- is +1 as O is −2 and so Cl + (−2) = −1, therefore Cl is +1.
- O in H_2O_2 (hydrogen peroxide) is −1, as each H is +1, so the two O atoms between them are −2.

Half equations for redox processes contain electrons.

- If a species is **reduced**, the electrons are on the **left**-hand side of the half equation.
- If a species is **oxidised**, the electrons are on the **right**-hand side of the half equation.
- The number of electrons is equal to the total change in oxidation number.
- For oxidising agents that contain oxygen (such as MnO_4^-), you will need H^+ on the left and H_2O on the right.

Some examples of half equations:

- for chlorine being reduced: $Cl_2 + 2e^- \rightarrow 2Cl^-$
- for manganate(VII) ions being reduced:

$$MnO_4^- + 8H^+ + 5e^- \rightarrow Mn^{2+} + 4H_2O$$

(The oxidation number of Mn changes by −5, so there have to be 5 electrons on the left.)

- for iodide ions being oxidised: $2I^- \rightarrow I_2 + 2e^-$

- **O**xidation **I**s **L**oss of electrons
- **R**eduction **I**s **G**ain of electrons
- Remember **OIL RIG**.

Overall redox equations can be obtained by adding half equations. You must check that:

- The number of electrons on the left of one half equation equals the number of electrons on the right of the other half equation.
- In the overall equation, the total change in oxidation number of the substance being reduced equals the total change in the substance being oxidised.
- The two reactants are on the left of the equation.

The overall equation for the oxidation of I^- ions by MnO_4^- ions is obtained from the two half equations:

$$MnO_4^- + 8H^+ + 5e^- \rightarrow Mn^{2+} + 4H_2O \quad \text{and} \quad 2I^- \rightarrow I_2 + 2e^-$$

The MnO_4^- half equation has 5 electrons, but the I^- equation only 2, so the MnO_4^- equation has to be multiplied by 2 and the I^- equation by 5, so that both have 10 electrons. They are then added to give

$$2MnO_4^- + 16H^+ + 10I^- \rightarrow 2Mn^{2+} + 8H_2O + 5I_2$$

The halogens

The boiling points of the halogens increase down the group.

- Fluorine and chlorine are gases, bromine is a volatile liquid, iodine is a solid. This is because the number of electrons in the halogen molecule increases, causing an increase in the strength of the intermolecular forces.

The oxidising power of the halogens decreases down the group.

- Chlorine will oxidise bromide ions but bromine will not oxidise chloride ions. Chlorine and bromine will oxidise Fe^{2+} ions but iodine will not. This is because the smaller chlorine atom has a greater attraction for electrons than the other halogens which are larger.

Chlorine (and bromine) reacts with water:

$$Cl_2 + H_2O \rightleftharpoons HCl + HOCl$$

Chlorine (and bromine) reacts with alkalis such as sodium hydroxide solution:

$$Cl_2 + 2OH^- \rightarrow Cl^- + OCl^- + H_2O$$

- In both reactions the chlorine has disproportionated, because it has been both oxidised (oxidation number goes up from 0 in Cl to +1 in HOCl) and reduced (oxidation number goes down from 0 in Cl to −1 in HCl).

Test for the halogens

- Chlorine is very pale green and bleaches damp litmus. It also turns colourless potassium bromide solution red-brown.
- Bromine is red-brown and slowly bleaches damp litmus.
- Iodine turns starch blue-black and solutions of iodine in hexane are violet.

The halides

Test for ionic halides (in solution):

Add dilute nitric acid and aqueous silver nitrate. Then add dilute ammonia, followed by concentrated ammonia to the precipitate that forms (this is a silver halide).

- Chlorides give a white ppt, bromides a pale cream ppt and iodides a pale yellow ppt
- The ppt of AgCl dissolves in dilute ammonia. The ppt of AgBr dissolves in conc ammonia but not dilute ammonia. The ppt of AgI does not dissolve.

Reducing power of halides

- Iodide ions are the strongest reducing agents and will reduce Fe^{3+} ions to Fe^{2+}.
- Bromide ions are a much weaker reducing agent, and chloride ions weaker still. Neither will reduce Fe^{3+} ions.
- When concentrated sulphuric acid is added to a solid ionic halide, gaseous hydrogen halide is first produced. This may then be oxidised to the halogen by the concentrated sulphuric acid, depending on the strength of the hydrogen halide as a reducing agent.
- When conc sulphuric acid is added to a chloride, steamy fumes of HCl are produced. This is too weak a reducing agent to be oxidised to chlorine by the sulphuric acid.
- When conc sulphuric acid is added to a solid bromide, misty fumes of HBr are produced some of which is oxidised by the sulphuric acid to red-brown bromine vapour. Some SO_2 is made by the reduction of the H_2SO_4.
- Hydrogen iodide is rapidly oxidised to violet fumes of iodine and reduces H_2SO_4 to H_2S and S.

Hydrogen halides as acids

- HCl, HBr and HI are all strong acids, but they are not as strong as sulphuric acid:

$$HCl(aq) + H_2O(l) \rightarrow H_3O^+(aq) + Cl^-(aq)$$

- Concentrated sulphuric acid will protonate halide ions to form the hydrogen halide, but see the box on 'Reducing power of halides' (left) for subsequent reactions.

Questions to try

Examiner's hints for question 1

(a) (i) Remember that the oxidation numbers in a polyatomic ion add up to the charge on the ion, and in a neutral substance they add up to zero.

(c) (i) An oxidising agent is reduced, and so it gains electrons. As the oxidising agent, FeO_4^-, is in acid solution, you will need to have H^+ ions on the left of the half equation. Remember that the number of electrons equals the change in oxidation number.

Q1

(a) (i) Define oxidation number (oxidation state).

....

....

[2 marks]

(ii) What is the oxidation number of chlorine in:

Cl_2 .... OCl^- .... $NaClO_3$ ....

[3 marks]

(b) (i) Write balanced equations for the reactions of chlorine with:

water .... [1 mark]

hydroxide ions .... [1 mark]

(ii) What, in terms of oxidation and reduction, has happened to the chlorine in these reactions?

....

....

[2 marks]

(c) Iron forms a highly oxidising ion of formula FeO_4^-.

(i) Write the half equation for its reduction to Fe^{3+} ions in acid solution.

....

[2 marks]

(ii) Write the half equation for the oxidation of Fe^{2+} ions to Fe^{3+} ions.

....

[1 mark]

(iii) Hence write the overall equation for the reaction between FeO_4^- ions and Fe^{2+} ions.

....

[2 marks]

[Total 14 marks]

Examiner's hints for question 2

(a) Remember that ionic halides react with concentrated sulphuric acid to produce gaseous hydrogen halides. These might then reduce the sulphuric acid.

(b) Halides in solution form a precipitate with aqueous silver ions. With chlorides, this precipitate will dissolve in dilute ammonia. With bromides, it is insoluble in dilute but soluble in concentrated ammonia. With iodides, it is insoluble in concentrated ammonia.

Q2

(a) The halide ions become stronger reducing agents down group VII. Use this fact to explain the reactions and observations when concentrated sulphuric acid is added to samples of potassium chloride, potassium bromide and potassium iodide.

....

....

....

....

....

[9 marks]

(b) A solid P is a mixture of the sodium salts of two different halides. A solution of P was acidified with dilute nitric acid and silver nitrate solution added. A creamy yellow precipitate W was obtained. Dilute ammonia solution was added to W and the mixture was filtered. The filtrate X was a colourless solution. This was made neutral by adding dilute nitric acid and a white precipitate Y was obtained. The residue Z was pale yellow and remained undissolved when concentrated ammonia solution was added.

(i) Identify the precipitates Y and Z.

Y is .... Z is .... [2 marks]

(ii) Hence identify **all** the ions in solid P.

....

[3 marks]

(iii) Write ionic equations to show the formation of W.

....

....

[2 marks]

The answers to these questions are on pages 83 and 84. [Total 16 marks]

Exam Questions and Student's Answers

1 (a) (i) Define standard enthalpy of combustion.

It is the energy change when 1 mole ✓ of a substance is burnt. ^^

[3 marks] 1/3

(ii) Write a balanced equations, including state symbols, for the reaction that represents the standard enthalpy of combustion of ethane, C_2H_6, at 25 °C.

$C_2H_6(g) + 3\frac{1}{2}O_2(g) \rightarrow$ ✓ $2CO_2(g) + 3H_2O(g)$ ✗

[2 marks] 

(b) Ethane can be made from ethene, $CH_2{=}CH_2$, by the reaction:

$$CH_2{=}CH_2(g) + H_2(g) \rightarrow CH_3{-}CH_3(g)$$

(i) Given the standard enthalpies of combustion below, draw a Hess's Law cycle and use it to calculate the standard enthalpy of this reaction.

Substance	ethane	ethene	hydrogen
ΔH_c^0/kJ mol^{-1}	–1560	–1409	–286

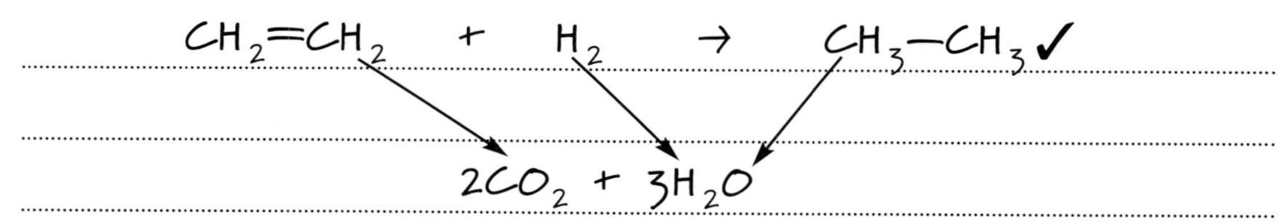

$\Delta H_r = \Sigma\ \Delta H$ of products $- \Sigma\ \Delta H$ of reactants ✗

$= -1560 - (-1409) - (-286) = +135$ kJ mol^{-1} ✗

[3 marks] 1/3

(ii) Draw an enthalpy level (profile) diagram for this reaction.

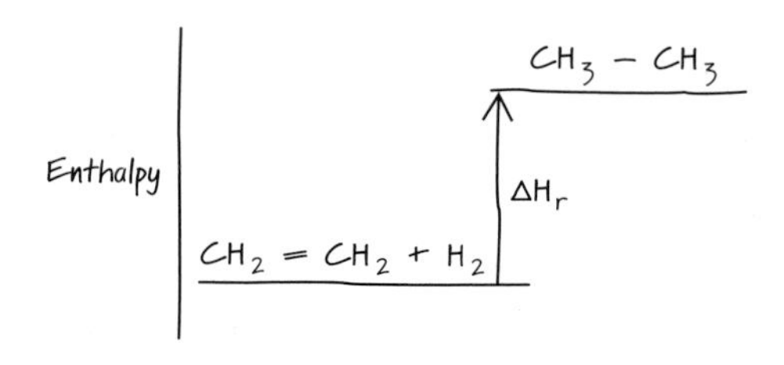

[1 mark] 1/1

(c) Ethane reacts with chlorine in a substitution reaction:

$$CH_3{-}CH_3(g) + Cl_2(g) \rightarrow CH_3{-}CH_2Cl + HCl \qquad \Delta H_r = -115\,kJ\,mol^{-1}$$

Given the bond enthalpies below, calculate the bond enthalpy of the C—Cl bond.

Bond	C—H	H—Cl	Cl—Cl
Average bond enthalpy /kJ mol^{-1}	412	431	242

ΔH_r = enthalpy change in bond making − enthalpy change in bond breaking ✗

Bonds broken 1 x C—H and 1 Cl—Cl

Bonds made 1 x C—Cl and 1 x H—Cl ✓

$\Delta H_{break} = 412 + 242 = 654$ $\qquad \Delta H_{make} = y + 431$ ✗

$-115 = y + 431 - 654$

$y = -115 - 431 + 654 = +108\,kJ\,mol^{-1}$ ✗

[4 marks]

2 When 4.75 g of zinc was added to 100 cm^3 of a 0.567 mol dm^{-3} solution of copper sulphate in an insulated polystyrene cup, the temperature rose from 16.3 °C to 45.7 °C.

The reaction is: $Zn(s) + CuSO_4(aq) \rightarrow Cu(s) + ZnSO_4(aq)$

Assume that the density of the solution is 1.00 g cm^{-3} and that its specific heat capacity is 4.18 J g^{-1} °C^{-1}, calculate:

(i) The moles of zinc and copper sulphate taken.

amount of zinc $= \dfrac{4.75\,g}{65.5\,g\,mol^{-1}} = 0.0725$ mol ✓

amount of copper sulphate $= 0.567\,mol\,dm^{-3} \times \dfrac{100}{1000}\,dm^3$

$= 0.0567$ mol ✓

[2 marks] 2/2

(ii) The heat produced in the experiment.

heat produced = mass x specific heat capacity x temperature change ✓

= 4.75 ✗ x 4.18 x 29.4 = 584 J

[2 marks] 1/2

(iii) ΔH for this reaction.

The zinc is in excess, so all the copper sulphate reacts. ✓

Heat produced per mole $= \frac{584}{0.0567}$ ✓ = 10295 J = 10.295 kJ

ΔH_r = ✗ 10.295 kJ mol^{-1}

[4 marks] 2/4

(iv) Why is this value not that of the standard enthalpy of reaction?

The solution used is not at the standard concentration ✓

of 1 mol dm^{-3}

[1 mark] 1/1

[Total 9 marks]

How to score full marks

1 (a) (i) **All enthalpy change definitions should refer to enthalpy change or heat change at constant pressure**. The correct answer is:

"The standard enthalpy of combustion is enthalpy change when **1 mole** of the substance is burnt in **excess oxygen** under **standard conditions**."

(ii) **Standard conditions are 1 atmosphere pressure (101 kPa) and a stated temperature, usually 25 °C (298K) as in this question**. Thus the water produced will be in the liquid state and so the equation should be:

$$C_2H_6(g) + 3\tfrac{1}{2}O_2(g) \rightarrow 2CO_2(g) + 3H_2O(l)$$

(b) (i) The expression $\Delta H_r = \Sigma \Delta H$ of products $- \Sigma \Delta H$ of reactants only works when using enthalpy of formation data. Here you must use the **Hess's Law cycle**, and so the correct calculation is:

$$CH_2{=}CH_2 + H_2\ (+3\tfrac{1}{2}O_2) \longrightarrow CH_3{-}CH_3\ (+3\tfrac{1}{2}O_2)$$

ΔH_c(ethane) $\quad \Delta H_c(H_2) \quad \Delta H_c$(ethane)

$$2CO_2 + 3H_2O$$

$\Delta H_r + \Delta H_c$(ethane) $= \Delta H_c$(ethene) $+ \Delta H_c$(hydrogen)

$\Delta H_r = (-1409) + (-286) - (-1560) = -135\ \text{kJ mol}^{-1}$

(c) The student made a common error here. Bond **breaking is endothermic** and bond **making is exothermic**. The correct answer is:

ΔH_r = enthalpy change in bond breaking + enthalpy change in bond making

Bonds broken	Bonds made
$1 \times$ C—H $= +412\ \text{kJ mol}^{-1}$	$1 \times$ C—Cl $= -y\ \text{kJ mol}^{-1}$
$1 \times$ Cl—Cl $= +242\ \text{kJ mol}^{-1}$	$1 \times$ H—Cl $= -431\ \text{kJ mol}^{-1}$

$-115 = (+412 + 242) + (-y - 431)$

$y = +115 + 412 + 242 - 431$

$= +338\ \text{kJ mol}^{-1}$

(ii) The mass in the heat produced expression is **the total mass of the reactants and the solvent**. In this case it is 4.75 + 100 = 104.75 g not 4.75 g.

(iii) The student correctly identified that it is the copper sulphate that reacts totally, and so divided the heat produced by the moles of copper sulphate to get the heat produced per mole. But then 2 marks were lost. The temperature increased which means that the reaction is **exothermic**, and so ΔH is **negative**.

ΔH = minus the heat produced per mole, and **the answer must not be given to 5 figures as only 3 are justified here**. The student's answer should have been:

$\Delta H_r = -10.3\ \text{kJ mol}^{-1}$

[If the student had correctly calculated the heat produced in (ii), ΔH_r would then equal $-227\ \text{kJ mol}^{-1}$, but the error of the wrong mass had already been penalised.]

Don't make these mistakes ...

Don't forget that all definitions refer to 1 **mole of substance**.

Don't forget to include **state symbols** in all thermochemical equations.

The 'formula' $\Delta H_r = \Sigma\Delta H$ of products $- \Sigma\Delta H$ of reactants is only true when using enthalpy of formation data, and, if used, must be written as $\Delta H_r = \Sigma\Delta H_f$ of products $- \Sigma\Delta H_f$ of reactants. **It is always better to draw and use a Hess's law cycle.**

Don't forget to learn the **definitions**.

In questions based on experimental data from a calorimeter, remember that the mass in the expression 'mass × specific heat capacity × ΔT is the **total** mass of the substances that heat up (or cool down). This includes the solvent.

Don't forget to **balance** a Hess's Law cycle.

Don't get the sign wrong. ΔH = minus the heat produced per mole. If the temperature rises, the reaction is exothermic and so ΔH is negative. If the temperature falls, the reaction is endothermic and ΔH is positive.

Key points to remember

Standard conditions must be mentioned in definitions. They are:

- A constant pressure of 101 kPa (1 atm)
- A stated temperature (usually 25 °C)
- All solutions at a concentration of 1 mol dm^{-3}

Exothermic reactions: the temperature rises, so heat is given out to the surroundings.

ΔH is negative.

Endothermic reactions: the temperature falls, so heat is taken in.

ΔH is positive

Standard enthalpy of formation is the enthalpy change when **1 mole** of a substance is formed from its **elements** under **standard conditions**. The standard enthalpy of formation of copper sulphate is the enthalpy change for the reaction:

$$Cu(s) + S(s) + 2O_2(g) \rightarrow CuSO_4(s)$$

Standard enthalpy of combustion is the enthalpy change when **1 mole** of a substance is burnt in **excess oxygen** under **standard conditions**. All ΔH_c values are **exothermic**. The standard enthalpy of combustion of ethanol, C_2H_5OH, is the enthalpy change for the reaction:

$$C_2H_5OH(l) + 3O_2(g) \rightarrow 2CO_2(g) + 3H_2O(l)$$

Standard enthalpy of neutralisation is the enthalpy change when an acid reacts with a base in 1 mol dm^{-3} solutions to **form 1 mole of water**. For a strong acid being neutralised by a strong base, it is the enthalpy change for the reaction:

$$H^+(aq) + OH^-(aq) \rightarrow H_2O(l)$$

Hess's Law states that the enthalpy change for a reaction is independent of the route by which the reaction is achieved.

In the cycle below:

$\Delta H = \Delta H_1 + \Delta H_2 + \Delta H_3$

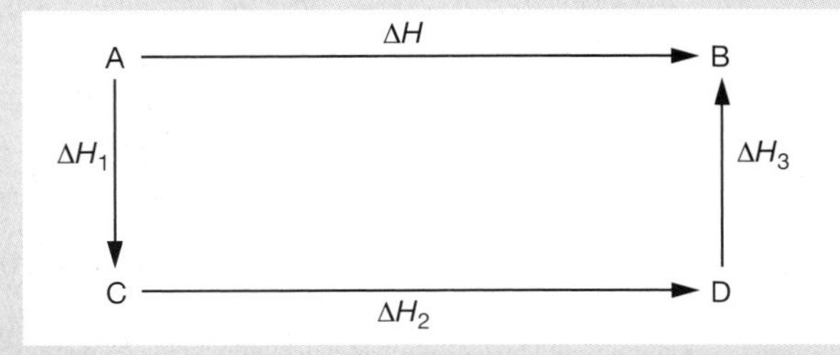

One common cycle is using **enthalpies of formation**:

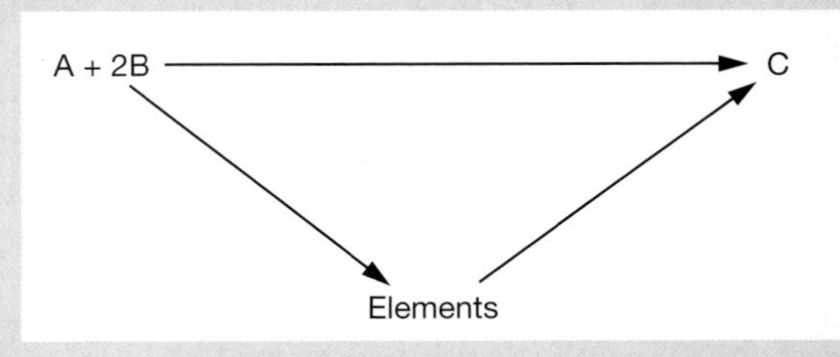

$\Delta H_r = -\Delta H_f$ of A $- 2\text{x}\ \Delta H_f$ of B $+ \Delta H_f$ of C

Another common cycle is using **enthalpies of combustion**

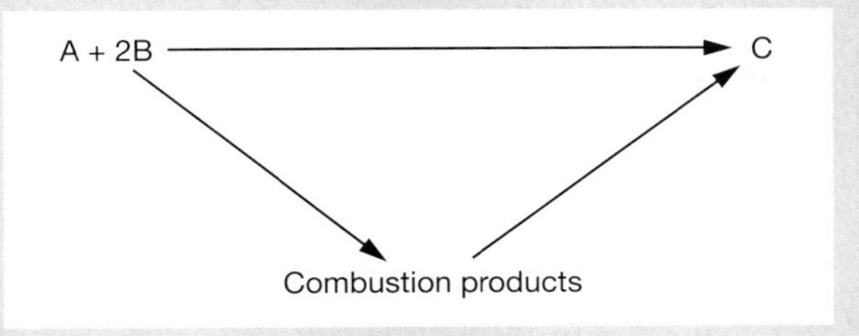

$\Delta H_r = \Delta H_c$ of A + 2x ΔH_c of B – ΔH_c of C

Calorimetry

Heat produced, q = m. c. ΔT

heat produced per mole = q/moles reacting

For **exo**thermic reactions ΔH is **negative**.

For **endo**thermic reactions ΔH is **positive**.

m = **total** mass of the contents of the calorimeter (solutions are assumed to have a mass equal to their volume).

c = the specific heat capacity (its value will be given in the question)

ΔT is the change in temperature.

Bond enthalpy is the enthalpy change per mole of bond broken. In calculations you need to use structural formulae to see what bonds are broken and what bonds are made.

- Bond breaking is endothermic (+)
- Bond making is exothermic (–)

ΔH reaction = enthalpy of bonds broken (a positive number)+ enthalpy of bonds made (a negative number).

Enthalpy level (profile) diagrams show the relative energy levels of the reactants (on the left) and of the products (on the right) with ΔH shown by an arrow. For the reaction

$A + B \rightarrow C \quad \Delta H = -\ 123\ \text{kJ mol}^{-1}$
The diagram is shown below:

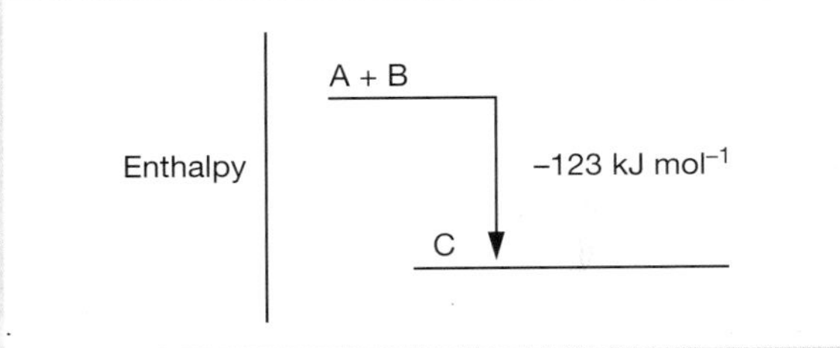

BOND BREAKING IS ENDOTHERMIC

BOND MAKING IS EXOTHERMIC

Questions to try

Examiner's hints for question 1

(a) (ii) It is sensible to add an equation with state symbols in all thermochemical equations.

(b) The alternative route in the cycle is via the elements.

(c) Draw structural formulae to see exactly what bonds are being broken and what bonds are being made.

Q1

(a) Define:

(i) standard conditions.

....

....

[3 marks]

(ii) the standard enthalpy of formation of a substance such as ethene, $CH_2{=}CH_2$.

....

....

[3 marks]

(b) Draw a Hess's Law cycle. Use it and the standard enthalpies of formation given in the table below to calculate the standard enthalpy of the reaction:

$$CH_2{=}CH_2(g) + H_2O(g) \rightarrow CH_3{-}CH_2OH(g)$$

	$CH_2{=}CH_2(g)$	$H_2O(g)$	$CH_3{-}CH_2OH(g)$
ΔH_f / kJ mol^{-1}	+ 53	– 242	– 235

[3 marks]

(c) (i) Use the average bond enthalpies given below to calculate the standard enthalpy change for the same reaction.

	C—H	C=C	H—O	C—C	C—O
Bond enthalpy / kJ mol^{-1}	412	612	463	348	360

[3 marks]

(ii) Explain why the answer to (c) (i) is different from that obtained in (b).

....

[2 marks]

[Total 14 marks]

Examiner's hints for question 2

(a) When using your graph to evaluate ΔT, remember that the reaction started at 3 minutes.

(b) What is the total mass that warmed up? Is the reaction exo- or endothermic?

(c) The aq in the equations stands for many moles of water, and so includes the $5H_2O$ that is in the final equation.

Q2

Water of mass 100 g was placed in a polystyrene cup and its temperature measured. After three minutes 4.80 g of anhydrous copper sulphate, $CuSO_4(s)$, was added and the mixture was thoroughly stirred for two minutes, by which time all the solid had dissolved. The temperature was measured for a further five minutes. The following results were obtained:

Temperature/°C	15.2	15.2	15.2	20.1	19.9	19.7	19.5	19.3	19.2
Time/minutes	0	1	2	5	6	7	8	9	10

Plot a graph of temperature (*y*-axis) against time (*x*-axis) and use it to determine the temperature change, ΔT.

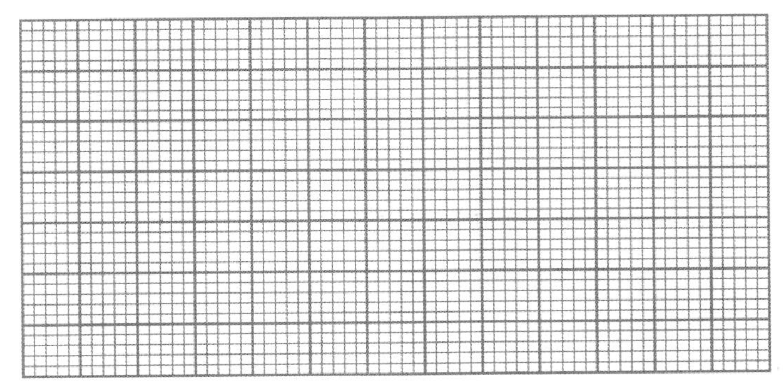

[5 marks]

(b) Hence calculate the molar enthalpy of solution of anhydrous copper sulphate

$$CuSO_4(s) + aq \rightarrow CuSO_4(aq)$$

given that the specific heat capacity of the solution is $4.18\ J\ g^{-1}\ °C^{-1}$ [4 marks]

(c) The molar enthalpy of solution of hydrated copper sulphate

$$CuSO_4.5H_2O(s) + aq \rightarrow CuSO_4(aq)$$

is $+4.7\ kJ\ mol^{-1}$. Draw a Hess's Law cycle or an energy level (profile) diagram and use it to calculate the enthalpy change for the reaction:

$$CuSO_4(s) + 5H_2O(l) \rightarrow CuSO_4.5H_2O(s)$$ [3 marks]

The answers to these questions are on pages 84, 85 and 86. [Total 12 marks]

Exam Questions and Student's Answers

1 The decomposition of an aqueous solution of hydrogen peroxide is slow at 40 °C but is faster at 50 °C.

(a) Explain, in terms of collision theory, the effect of this temperature increase on the rate of this reaction.

When the temperature is increased the atoms ✗ move faster as they have more kinetic energy. This means that they collide more often ✓ and with more force. There are more collisions and more of the colliding molecules have energy that is greater than the activation energy ✓ and so there are more effective ✗ collisions.

[4 marks] 2/4

(b) This decomposition is catalysed by a solution of bromine. Explain, in terms of collision theory, why reactions are speeded up by catalysts.

A catalyst lowers ✓ the activation energy of the reaction ^ and so a greater proportion of the molecules have the activation energy and therefore will react on collision. ✓

[3 marks] 2/3

(c) If a more concentrated solution of hydrogen peroxide is used, the rate of decomposition at 50 °C will be different. State and explain, in terms of collision theory, the difference that altering the concentration will make on the rate.

^ In a more concentrated solution the molecules are closer ✓ together and so will collide more frequently. ✓

[3 marks] 2/3

[Total 10 marks]

2 The distribution of molecular energies at a temperature T_1 is shown below.

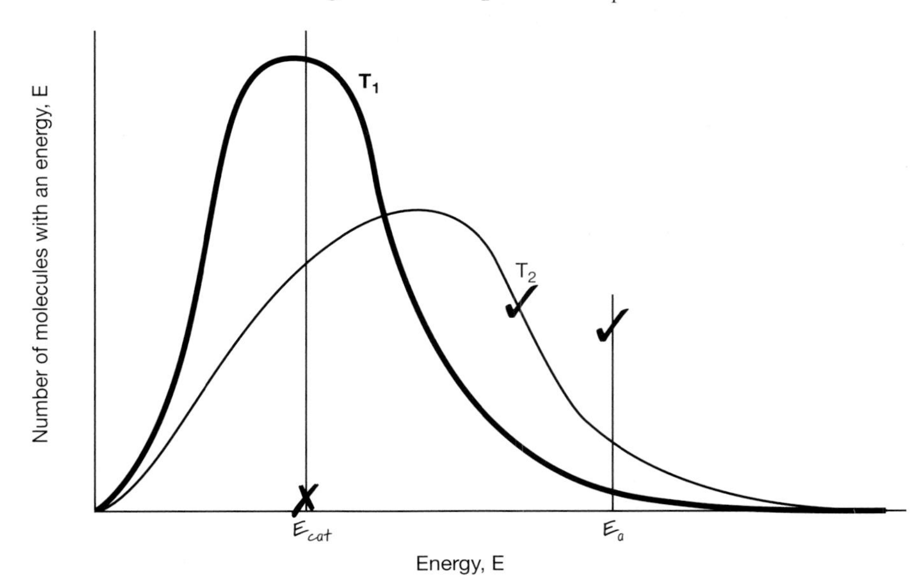

(a) Sketch a graph, labelling it T_2, showing the distribution of molecular energies at a higher temperature T_2 and hence explain why the rate of reaction is greater at temperature T_2 than at T_1.

At T_2 there are more molecules that have energy greater ✓ than the activation energy, E_a. ∧∧

[5 marks] 

(b) Explain the effect of a catalyst on the rate of a reaction. You should use the graph in your answer.

∧ The activation energy for the catalysed route is lower ✓ than that for the uncatalysed reaction. ∧

[4 marks] 

[Total 9 marks]

How to score full marks

(a) Marks were lost because:

- the student used the word **atom** when describing a substance that is **molecular**.
- the phrase "more effective collisions" is ambiguous. **You should write that "more of the collisions result in reaction".**

The student gained a mark for stating that more of the colliding molecules have energy greater than the activation energy.

Although the student gained a mark for stating that there are more collisions, it would have been better to say that **the frequency of collisions is increased**.

(b) **The activation energy of a reaction cannot be altered.** A catalyst provides an alternative path or reaction route, which has a lower activation energy.

(c) The question asked you to state and explain. This answer does not state what happens to the rate of reaction. Your answer should begin; "The rate of reaction is increased because in a more..."

2 **(a)** Marks were lost because:

- the student should have stated **why** the graph shows that more molecules have energy greater than the activation energy. The correct answer is **because the area under the graph to the right of the E_a line is greater for the T_2 graph than for the T_1 graph.**

- the student should also have said that this means that **more of the collisions will result in reaction** (or more of the collisions will be successful).

The student scored the first mark, which is for the shape of the graph for T_2. The peak for T_2 **must be to the right and lower than the peak for T_1** and **the area under the two curves should be the same**, but if you draw it badly write "The areas under the two graphs should be the same", **then you will not be penalised**.

The student gained the second mark because the activation energy for the uncatalysed reaction was correctly drawn to the right of the peaks at both temperatures.

The third mark was gained for saying that more molecules have energy greater than the activation energy.

(b) The student lost a mark because the line for the catalysed activation energy, E_{cat}, was drawn in the wrong place on the graph. **The line must be to the right of both peaks.**

The student failed to state that **the rate of reaction would be faster with a catalyst**.

The student should also have explained that the **lower activation energy** means that a **greater proportion of the molecules** will have the necessary energy to **react on collision**.

Don't make these mistakes ...

Don't state that increasing the temperature brings about an increase in the rate of a reaction by causing a greater collision frequency. A temperature increase of 10 °C causes only a very slight increase in the frequency of collision (about a 1% increase). **The important point is that this temperature rise causes a 100% increase in the fraction of molecules that have the activation energy to react on collision**.

Don't say that a catalyst speeds up a reaction by lowering the activation energy. Homogeneous catalysts (those that are in the same state as the reactants, e.g. Fe^{2+} ions in the oxidation of iodide ions by persulphate ions) work by combining with one of the reactants to form an intermediate, which then reacts to reform the catalyst and the products. **This alternate reaction route has a lower activation energy**.

Heterogeneous catalysts (those that are in a different state to the reactants, e.g. iron in the Haber process for the manufacture of ammonia from nitrogen and hydrogen gases) adsorb the reactants on their surface. The adsorbed gases react to form the products and then leave the surface of the catalyst. **This alternate reaction route also has a lower activation energy**.

Key points to remember

Maxwell-Boltzmann distribution

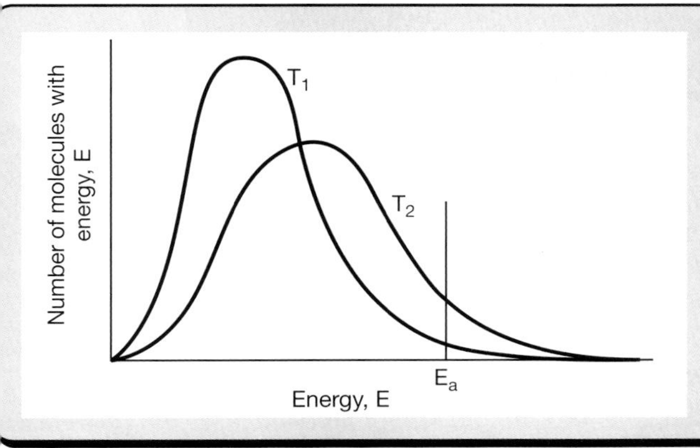

This shows the way in which the kinetic energy of the molecules in a gas (or liquid) are distributed at a temperature T_1. The number of molecules that have energies greater than or equal to a value E_a is equal to the area under the curve to the right of the E_a line.

At a higher temperature T_2, a larger number have energies greater than or equal to this value.

Factors that affect the rate of reaction

Factor	How it affects the rate	Why it affects the reaction
Temperature	An increase in temperature increases the rate of a chemical reaction. This is true for both exothermic and endothermic reactions.	The particles have more kinetic energy so more of them have energy ≥ the activation energy. Thus more of them react on collision. The collision frequency also increases but slightly (see diagram above).
Concentration	An increase in the concentration of substances in solution increases the rate of reaction.	The particles are closer together and so the frequency of collision is increased.
Pressure	An increase in pressure of a gas increases the rate of reaction.	As above, the gas molecules are closer together and so the frequency of collision is increased.
Catalyst	A catalyst speeds up a reaction without getting used up.	A catalyst provides an alternative reaction route which has a lower activation energy, thus , at a given temperature, more particles will have kinetic energy greater than or equal to the activation energy.
Surface area	The rate of a reaction of a solid with a gas or a liquid is dependent on the surface area of the solid – the bigger the surface area the faster the reaction.	The frequency of collision of a reactant molecule or ion with the surface is increased.

Questions to try

Examiner's hint for question 1
There are 5 factors which may affect the rate of a reaction. Which are applicable in each reaction?

Q1

(a) An aqueous solution of 1-chloropropane reacts slowly with dilute sodium hydroxide at 25 °C according to the equation:

$$CH_3CH_2CH_2Cl\ (aq) + OH^-(aq) \rightarrow CH_3CH_2CH_2OH(aq) + Cl^-(aq)$$

State three ways by which the rate of this reaction could be increased.

....

....

....

[3 marks]

(b) Explain why the flour dust in a flour milling factory may explode, whereas it is difficult to ignite ordinary flour in the home.

....

....

....

[3 marks]

(c) Consider the reaction: $Zn(s) + 2HCl(aq) \rightarrow ZnCl_2(aq) + H_2(g)$

Explain the effect, if any, on the rate of reaction of:

(i) Increasing the pressure

....

[2 marks]

(ii) Adding water

....

[2 marks]

(iii) Using hydrobromic acid (HBr), of the same concentration, instead of hydrochloric acid.

....

....

[2 marks]

[Total 12 marks]

Examiner's hint for question 2
For questions involving the Maxwell-Boltzmann distribution, you may need to draw two graphs at different temperatures. These must be clearly labelled. Draw in a suitable activation energy and use collision theory to **state** and **explain** the effect on the rate.

Q2

(a) Define activation energy.

....

....

[2 marks]

(b) Nitrogen(V) oxide decomposes slowly at room temperature according to the equation:

$$2N_2O_5(g) \rightarrow 4NO_2(g) + O_2(g)$$

(i) Draw Maxwell-Boltzmann distribution of energies graphs at two different temperatures, and use them to explain how an increase of temperature affects the rate of the reaction.

Number of molecules with energy, E

Energy, E

....

....

....

....

[8 marks]

(ii) Use collision theory to explain how the rate of this reaction would be altered by an increase in pressure.

....

....

....

[3 marks]

The answers to these questions are on pages 86 and 87. [Total 13 marks]

Exam Question and Student's Answer

1 **(a)** What is meant by the term "dynamic equilibrium"?

The system reacts rapidly until the substances reach equilibrium, when there is no further reaction. ✗

[2 marks] 0/2

(b) In the Haber process hydrogen is manufactured from methane, air and water. The first step in this process is the reaction between methane, CH_4, and steam at a temperature of 1000 K in the presence of a nickel catalyst at a pressure of 30 atmospheres.

$$CH_4(g) + H_2O(g) \rightleftharpoons CO(g) + 3H_2(g) \qquad \Delta H = +\,206\,\text{kJ mol}^{-1}$$

State and explain the effect on the position of equilibrium of the pressure being reduced to a lower value at the same temperature.

A decrease in pressure will drive the reaction ^ to the side with more ^ molecules.

[2 marks] 0/2

(c) In the second step the carbon monoxide is oxidised by water at 700 K in the presence of an iron oxide catalyst.

$$CO(g) + H_2O(g) \rightleftharpoons CO_2(g) + H_2(g) \qquad \Delta H = -\,41\,\text{kJ mol}^{-1}$$

(i) State and explain the effect on the position of equilibrium of increasing the temperature at constant pressure.

An increase in temperature will drive the equilibrium to the left, ✓ as the reaction is exothermic ^.

[2 marks] 1/2

(ii) Why is a temperature of 700 K used rather than a higher or a lower value?

If the temperature were higher the yield would be less, ✓ but if the temperature were lower, although the yield would be greater, ✓ the rate of reaction would be much less ✓ ^.

[4 marks] 3/4

(iii) Explain why a catalyst is used in this manufacturing process.

A catalyst increases the rate of reaching equilibrium, and so allows the reaction to take place quickly at a temperature, which is not too high and expensive, and at which there is a reasonable yield.

[3 marks] 

(d) Hydrated cobalt(II) ions, which are pink, react reversibly with a solution of chloride ions:

$$\underset{\text{pink}}{[Co(H_2O)_6]^{2+}} + 4Cl^- \rightleftharpoons \underset{\text{blue}}{[CoCl_4]^{2-}} + 6H_2O$$

When a solution containing blue $[CoCl_4]^{2-}$ ions is cooled in iced water, the solution turns pink. Deduce whether the reaction as written above is exo or endothermic.

A decrease in temperature drives the reaction to the left, and so this must be the endothermic direction. Therefore the reaction as written is endothermic left to right.

9/16

[3 marks] 

[Total 16 marks]

How to score full marks

(a) The student's answer is wrong. **When a question says 'what is meant by...' you must explain both words – dynamic and equilibrium.** The correct answer is:

"Dynamic equilibrium is when the rate of the forward reaction equals the rate of the reverse reaction, and so there is no further change in the concentrations of any of the substances in the equilibrium mixture."

(b) The student is asked to state and explain the effect of a pressure decrease but fails to do this. **The effect is that the position of equilibrium is driven to the right, and the explanation is that the right-hand side contains more gas molecules than the left.** (This can be predicted by Le Chatelier's principle – see Key Points on page 66.)

(c) **(i)** The student correctly stated the effect of increasing the temperature, but the explanation was slightly ambiguous, so only 1 mark was given. **An equilibrium reaction goes in both directions, so it is important to make it clear which direction is exothermic and which endothermic.** The correct answer is:

"The position of equilibrium shifts to the left, because an increase of temperature will drive the equilibrium in the endothermic direction, which is from right to left."

(ii) **There are four marks for this question, but the student only made three points** and so scored 3 marks. The student missed out the important point that the temperature of 700 K is a compromise temperature giving a reasonable yield at a good rate.

(iii) The student omitted the vital fact that **a catalyst has <u>no</u> effect on the <u>position</u> of equilibrium.**

Key points to remember

Dynamic equilibrium

There is no further change in the concentrations of any of the chemicals, because the rate of the forward reaction equals the rate of the reverse reaction.

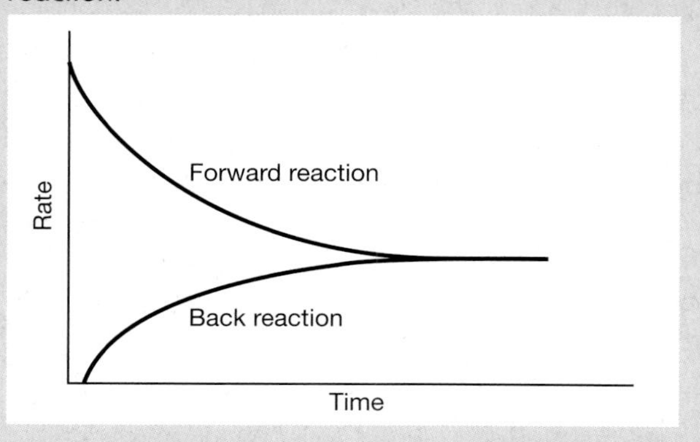

Le Chatelier's principle is: "If the conditions of a system at equilibrium are altered, the system will react as if to restore the original conditions".

This principle can be used to predict changes in position of equilibrium. It does not explain them.

When a reaction is in equilibrium, an **increase in temperature** will:

- move the position of equilibrium in the **endothermic** direction.
- increase the rate of the forward *and* the back reactions. However, the rate of the endothermic reaction is increased more than the rate of the exothermic reaction.
- mean that equilibrium is reached sooner at higher temperatures.

An example:

$N_2(g) + 3H_2(g) \rightleftharpoons 2NH_3(g) \quad \Delta H = -92\,kJ\,mol^{-1}$

- The reaction is exothermic left to right.
- A temperature increase causes the position of equilibrium to move to the left.
- A temperature decrease causes the position of equilibrium to move to the right, but will make the reaction slower.
- A compromise temperature of 400 °C is used in the manufacture of ammonia.

- When a reaction is in equilibrium, an **increase in pressure** will move the position of equilibrium to the side with **fewer gas molecules**.

  An example:
  $N_2(g) + 3H_2(g) \rightleftharpoons 2NH_3(g)$
  4 gas molecules $\rightleftharpoons$ 2 gas molecules
- A high pressure increases the rate (slightly).
- A very high pressure is very expensive.
- An increase in pressure will move the position of equilibrium to the right. As the yield at ordinary pressures is very low, a pressure of 200 atmospheres is used.

The effect of a catalyst

A catalyst has no effect on the position of equilibrium; it speeds up the rate of the forward and the back reactions equally, and so equilibrium is reached more quickly.

- The catalyst for the Haber process for manufacturing ammonia is iron.
- The catalyst for the contact process for manufacturing sulphuric acid is vanadium(V) oxide.

Removal of a reactant

If a reactant is removed physically or chemically, the position of equilibrium will shift to make more of that reactant.

If sodium hydroxide solution is added to the equilibrium:

$$CH_3COOC_2H_5 + H_2O \rightleftharpoons CH_3COOH + C_2H_5OH$$

it removes the ethanoic acid, CH_3COOH, and drives the equilibrium to the right.

Increase in concentration of a reactant

If the concentration of a reactant is increased, the position of equilibrium will shift to remove that reactant.

If concentrated HCl is added to the equilibrium:

$$PbCl_2(s) + 2Cl^-(aq) \rightleftharpoons PbCl_4^{2-}(aq)$$

the concentration of Cl^- ions is increased and so the equilibrium is shifted to the right. Thus the solid $PbCl_2$ will dissolve.

Economic factors

The aim of an industrial process is a high yield made quickly and cheaply.

- A high pressure is very expensive.
- A very high temperature is expensive.
- Unused reactants are recycled through the reaction vessel.
- Exothermic processes are carried out at a temperature that is a compromise between yield and rate.
- A catalyst enables the reaction to proceed at a temperature that gives a good yield quickly.
- A high pressure (other than that required to push the gases through the plant) is only used if the yield is otherwise very low.

Don't make these mistakes ...

Don't think that **all reactions cease** when equilibrium has been reached.

Don't forget that **equilibrium can only be reached in a closed system**, which is one in which no chemicals enter or leave the reaction mixture.

Don't forget to discuss rate as well as yield (or position of equilibrium) when answering a question on the optimum industrial conditions.

Don't muddle pressure with volume. For equilibrium reactions that involve gases, an increase in volume will cause a decrease in pressure, which will then result in the position of equilibrium being moved to the side with more gas molecules.

Question to try

Examiner's hints
1 (c) and **(d)**
Remember that Le Chatelier's principle predicts what happens if the conditions of a system in equilibrium are altered.
High pressure is expensive and is only used for low yield reactions.

Q1

(a) Gases A and B were mixed and allowed to reach equilibrium in a closed container:

$$A(g) + B(g) \rightleftharpoons C(g)$$

Draw on the axes below how the rates of the forward and back reactions alter over time until equilibrium is reached.

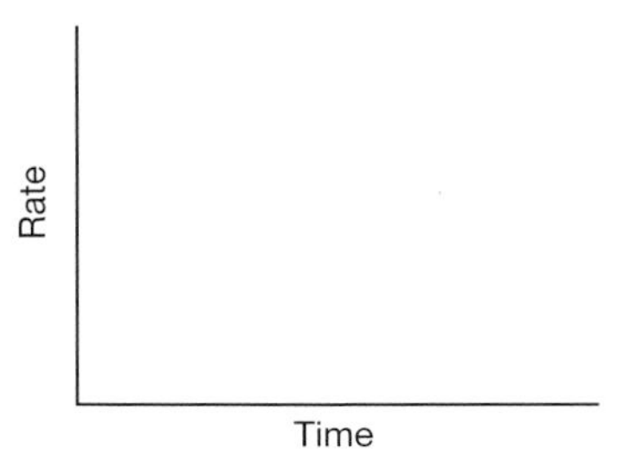

[3 marks]

(b) Which statements about a system in equilibrium are true and which are false?

No reactions are occurring. ....

The concentrations of the substances do not alter. ....

The amount of product equals the amount of reactant. ....

The rate of the forward reaction equals the rate of the reverse (back) reaction. ....

[4 marks]

(c) The critical step in the manufacture of sulphuric acid is the oxidation of sulphur dioxide.

$$SO_2(g) + \tfrac{1}{2}O_2(g) \rightleftharpoons SO_3(g) \qquad \Delta H = -98\ \text{kJ mol}^{-1}$$

Explain why the following conditions are chosen to make the process as economic as possible.

(i) A temperature of 425 °C

....

....

....

[4 marks]

(ii) A pressure of 202 kPa (2 atm)

....

....

[2 marks]

(iii) A catalyst of vanadium(V) oxide

....

....

....

[3 marks]

(d) Yellow chromate(VI) ions in aqueous solution are in equilibrium with orange dichromate(VI) ions.

$$2CrO_4^{2-}(aq) + 2H^+(aq) \rightleftharpoons Cr_2O_7^{2-}(aq) + H_2O(l)$$

State and explain what you would see when:

(i) a concentrated solution of sodium hydroxide is added to a solution containing dichromate(VI) ions.

....

....

[3 marks]

(ii) dilute sulphuric acid is added to a solution of chromate(VI) ions.

....

....

[3 marks]

The answers to this question are on pages 87 and 88. [Total 22 marks]

Exam Questions and Student's Answers

1 (a) Alkenes often react by electrophilic addition.

(i) Give an example of such a reaction, naming the reagent and stating the equation.

Reagent: Bromine ✓

Equation: $CH_2{=}CH_2 + Br_2 \rightarrow C_2H_4Br_2$ ✗ [2 marks] 1/2

(ii) Explain the meaning of the term *electrophile*.

It is a δ+ species which is attracted to a negative or δ– site ✗

[2 marks] 0/2

(iii) Explain the meaning of the term *addition reaction*.

It is a reaction when two substances add together. [1 mark] 0/1

(b) Consider the following reaction scheme:

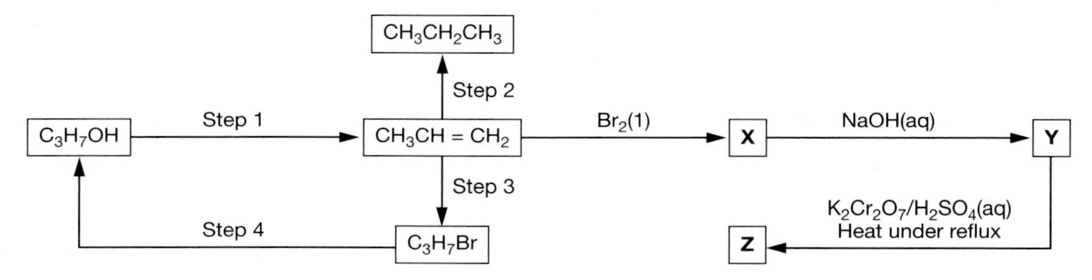

(i) Identify **X**, **Y** and **Z**.

X $CH_3CHBrCH_2Br$ ✓

Y $CH_3CH(OH)CH_2(OH)$ ✓

Z $CH_3CH(OH)C(=O)OH$ ✓ [4 marks] 3/4

(ii) Name **X**.

1,2-dibromopropane ✓ [1 mark] 1/1

(iii) State the reagents and conditions for steps 1, 2, 3 and 4.

Step 1: sulphuric acid at 170°C ✓

Step 2: hydrogen ✓ a heated nickel catalyst ✓

Step 3: hydrogen bromide ✓ aqueous solution ✗

Step 4: sodium hydroxide heat [8 marks] 4/8

(iv) State the structural formula of C_3H_7Br which is formed in step 3, and explain why this compound is the major product.

$CH_3CHBrCH_3$ ✓ because Markovnikov's rule states that the hydrogen goes on to the carbon atom which already has more hydrogens attached to it. ✗

[3 marks] 

(c) Vinyl alcohol, $CH_2{=}CHOH$, can be polymerised. Draw a section of its polymer showing two repeat units.

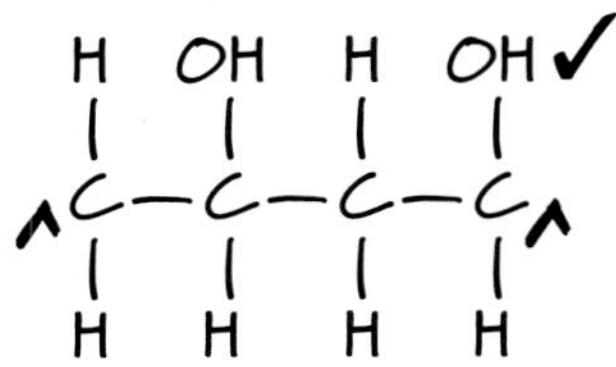

[2 marks] 

[Total 23 marks]

2 (not for Edexcel candidates)

(a) (i) Outline, using curly arrows, the mechanism for the reaction between aqueous hydroxide ions and iodoethane.

OH^- ✗ + CH_3, H, H–C–I ✓ → [H–O····C····I]$^-$ → H–O–C(CH_3)(H)–H ✓ + I^-

[4 marks] 

(ii) Classify this mechanism.

It is a nucleophilic ✓ substitution ✓ reaction.

[2 marks] 

(b) Ammonia reacts with iodoethane by a similar mechanism. What, in the ammonia molecule, makes this possible?

The nitrogen atom is very δ−. ✗

[1 mark] 0/1

[Total 7 marks]

How to score full marks

1 (a) (i) The formula given by the student – $C_2H_4Br_2$ – is ambiguous. The correct product is CH_2BrCH_2Br,not its isomer CH_3CHBr_2. **Never use 'added up' formulae if there is doubt as to the exact structure.**

The critical points you need to make to explain the meaning of electrophile are:

- It accepts a lone pair of electrons to form a covalent bond;
- It attacks regions of high electron density (usually π bonds).

(iii) When defining addition, don't use the word 'add'. The correct answer is:

"An addition reaction is when two substances react together to form a **single** product."

(b) (i) Compound Y has a primary and a secondary OH group. Both will be oxidised, so Z is $CH_3COCOOH$ (a ketoacid).

$$CH_3—C(=O)—C(=O)—O—H$$

(iii) Step 1: You need to write that the reagent is concentrated sulphuric acid.

Step 3: The hydrogen bromide, HBr, must be gaseous. The correct condition is 'room temperature'.

Step 4: You need to write that the reagent is dilute (aqueous) sodium hydroxide and that the conditions are 'heat under reflux' (as the bromoethane is very volatile).

(iv) Markovnikov's rule does not explain the direction of addition, it only tells you the product. The correct answer is:

"If the hydrogen atom adds to the terminal (end) carbon atom, the **secondary carbocation** formed will be **more stable than the primary carbocation** which is the intermediate for the other isomer. **This is because of the electron 'pushing' effect of the CH_3 groups which reduce the positive charge.**"

($CH_3CH^+CH_3$ is more stable than $CH_3CH_2CH_2^+$)

(c) The candidate has left out the 'continuation' bonds. The correct answer is:

$$\left(\begin{array}{cccc} H & OH & H & OH \\ | & | & | & | \\ —C— & C— & C— & C— \\ | & | & | & | \\ H & H & H & H \end{array} \right)$$

2 (a) (i) There are two errors in the mechanism. The curly arrow **must start from the lone pair of the <u>oxygen</u> atom in the OH^- ion**. The second error is that the student **omitted the negative charge on the transition state** (intermediate). The correct mechanism is:

$$H—\ddot{O}^{\ominus} + C(CH_3)(H)(H)—I \rightarrow \left[H—O\cdots C(CH_3)(H)(H)\cdots I \right]^{\ominus} \rightarrow H—O—C(CH_3)(H)—H + I^{\ominus}$$

(b) Nucleophiles must have a lone pair of electrons which they use to form a covalent bond (in this case with the carbon atom). The correct answer is:

"The nitrogen atom in ammonia has a **lone pair of electrons** as does the oxygen in the hydroxide ion."

Don't make these mistakes ...

Don't forget to define nucleophiles and electrophiles in terms of donating or accepting a lone pair of electrons and hence forming a covalent bond.

When describing the test for alkenes, don't forget to give the **colour** of the bromine solution **before** and **after** the reaction.

In mechanisms, ensure that a **curly arrow** starts from a bond or a lone pair, and ends forming an anion or a bond.

Always write cis/trans isomers with the correct **120°** bond angle around the double bonded carbon atoms.

Don't forget that **primary alcohols** are oxidised by dilute sulphuric acid and potassium dichromate(VI) first to an aldehyde and then to a carboxylic acid. **Secondary alcohols** are oxidised to ketones and **tertiary alcohols** are not oxidised.

Alcohols don't react with nucleophiles such as CN^- ions (from KCN)

Don't use 'added up' formulae such as C_3H_7OH; this could be either propan-1-ol or propan-2-ol.

The product of a halogenoalkane with potassium (or sodium) hydroxide **depends on the solvent**. Aqueous conditions give an alcohol (nucleophilic substitution), ethanolic solutions an alkene (elimination).

Key points to remember

Definitions

- **Addition reaction**: a reaction in which two substances combine to form a **single** product.
- **Substitution reaction**: a reaction in which one atom or group replaces another in a molecule.
- **Elimination reaction**: a reaction in which a simple molecule, such as HBr or H_2O, is removed from a larger one and is not replaced by any other atom or group. The result is a compound with a double bond.
- **Free radical**: a species with an unpaired electron
- **Nucleophile**: a species with a lone pair of electrons, which it donates to form a covalent bond. It is attracted to electron deficient sites
- **Electrophile**: a species that accepts a lone pair of electrons and forms a new covalent bond. It is attracted to electron rich sites

Reactions of alkanes e.g. ethane

Reaction	Conditions
$+O_2$: $2C_2H_6 + 7O_2 \rightarrow 4CO_2 + 6H_2O$	Heat or spark
$+Cl_2$(or Br_2): $C_2H_6 + Cl_2 \rightarrow C_2H_5Cl + HCl$	UV light or sunlight

Key points to remember

Summary of the reactions of alkenes

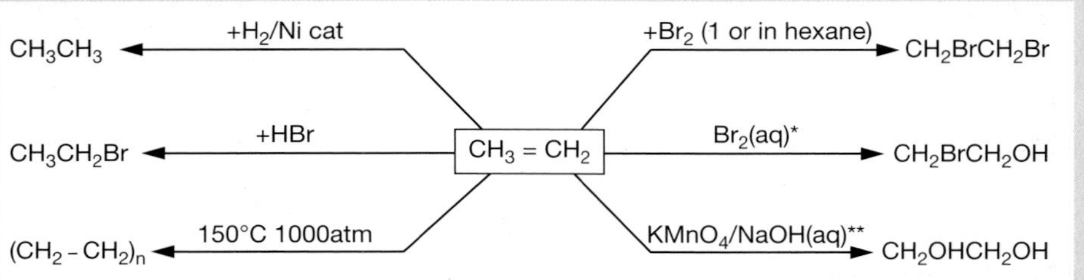

* Not required by Edexcel

** Only required by OCR

Reactions of alkenes e.g. ethene

Reaction	Conditions/observations
$+Br_2(l)$ or in hexane $CH_2{=}CH_2 + Br_2 \rightarrow CH_2BrCH_2Br$	Red-brown bromine turns colourless
$+Br_2$ (aq) $CH_2{=}CH_2 + Br_2 + H_2O \rightarrow CH_2BrCH_2OH + HBr$	Red-brown bromine water turns colourless
+HBr (or HCl or HI) $CH_2{=}CH_2 + HBr \rightarrow CH_3CH_2Br$	Mix gases at room temperature
$+H_2$ $CH_2{=}CH_2 + H_2 \rightarrow CH_3CH_3$	Pass gases over nickel catalyst at 100 °C or over platinum catalyst at room temperature
$+H_2O$ $CH_2{=}CH_2 + H_2O(g) \rightarrow CH_3CH_2OH$	Pass ethene and steam over a phosphoric(V) acid catalyst on silica at 300 °C and a pressure of 60 atm.
polymerisation $n\ CH_2{=}CH_2 \rightarrow (CH_2{-}CH_2)_n$	Either: 1000 atm pressure, 150 °C and a trace of oxygen. Or: triethyl aluminium and titanium tetrachloride catalysts

Reactions of alcohols

Reaction	Conditions / observations
+ potassium dichromate(VI) and dilute sulphuric acid	Heat under reflux
$CH_3CH_2OH + 2[O] \rightarrow CH_3COOH + H_2O$	Primary alcohols are oxidised via aldehydes to carboxylic acids and turn the dichromate green
$CH_3CH(OH)CH_3 + [O] \rightarrow CH_3COCH_3 + H_2O$	Secondary alcohols are oxidised to ketones and also turn the dichromate green
$(CH_3)_3COH \rightarrow$ no reaction	Tertiary alcohols are not oxidised and the dichromate stays orange
+ concentrated sulphuric acid $CH_3CH_2OH - H_2O \rightarrow CH_2{=}CH_2$	Heat to 170 °C with excess sulphuric acid and the alcohol will be dehydrated (or pass ethanol vapour over hot Al_2O_3)

Alcohols

- **Primary alcohols** have one carbon atom directly bonded to the carbon of the C—OH group:

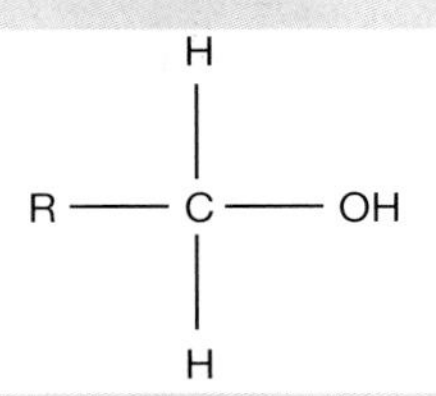

- **Secondary alcohols** have two carbon atoms directly bonded to the carbon of the C—OH group:

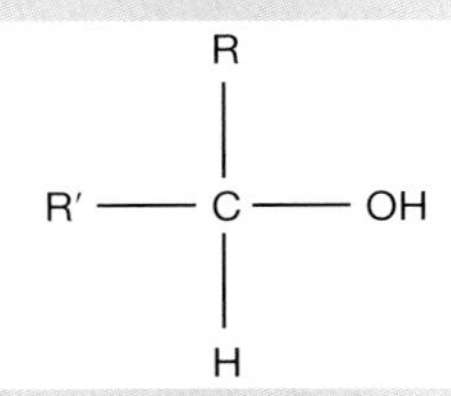

- **Tertiary alcohols** have three carbon atoms directly bonded to the carbon of the C—OH group:

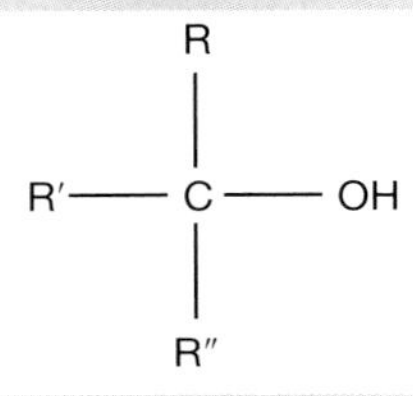

Summary of the reactions of alcohols

CH_3CH_2OH

- Na → $(CH_3CH_2O^-Na^+)$ + $H_2(g)$**
- CH_3COOH, conc H_2SO_4 → $CH_3COOCH_2CH_3$**
- HBr (from KBr(s) & conc H_2SO_4) → CH_3CH_2Br***
- $K_2Cr_2O_7/H_2SO_4(aq)$ → CH_3COOH
- PCl_5 → CH_3CH_2Cl + HCl(g)*
- conc H_2SO_4 at 170°C → $CH_2 = CH_2$

* Edexcel only ** OCR only
*** Edexcel and OCR

Summary of the reactions of halogenoalkanes

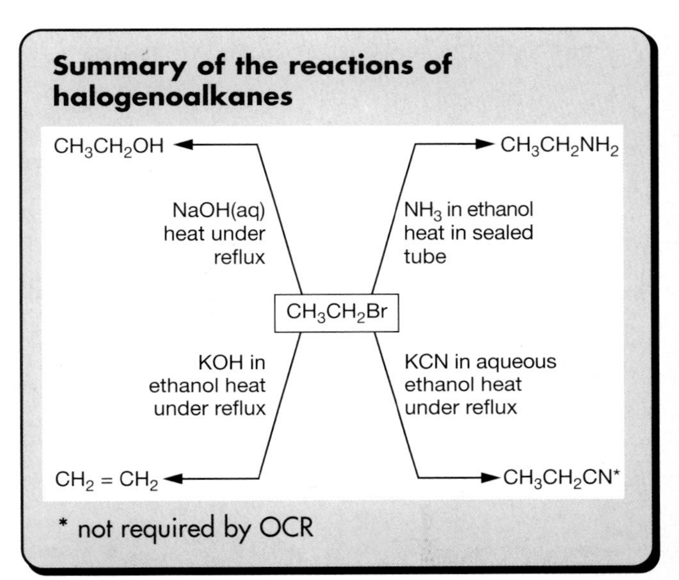

* not required by OCR

Reactions of halogenoalkanes (haloalkanes)

Reaction	**Conditions**
+ **aqueous** NaOH (or KOH) $CH_3CH_2Br + OH^- \rightarrow CH_3CH_2OH + Br^-$	Heat under reflux in aqueous solution. (a substitution reaction)
+ KOH dissolved in **ethanol** $CH_3CH_2Br + KOH \rightarrow CH_2{=}CH_2 + H_2O + KBr$	Heat under reflux in ethanolic solution (an elimination reaction)
+ NH_3 $CH_3CH_2Br + NH_3 \rightarrow CH_3CH_2NH_2 + HBr$ or to the salt $\rightarrow CH_3CH_2NH_3^+Br^-$	Heat the ammonia dissolved in ethanol with the bromoethane in a sealed tube
+ KCN (not required for OCR candidates) $CH_3CH_2Br + KCN \rightarrow CH_3CH_2CN + KBr$	Heat under reflux in an aqueous ethanolic solution

Questions to try

Q1

Consider the reaction scheme below:

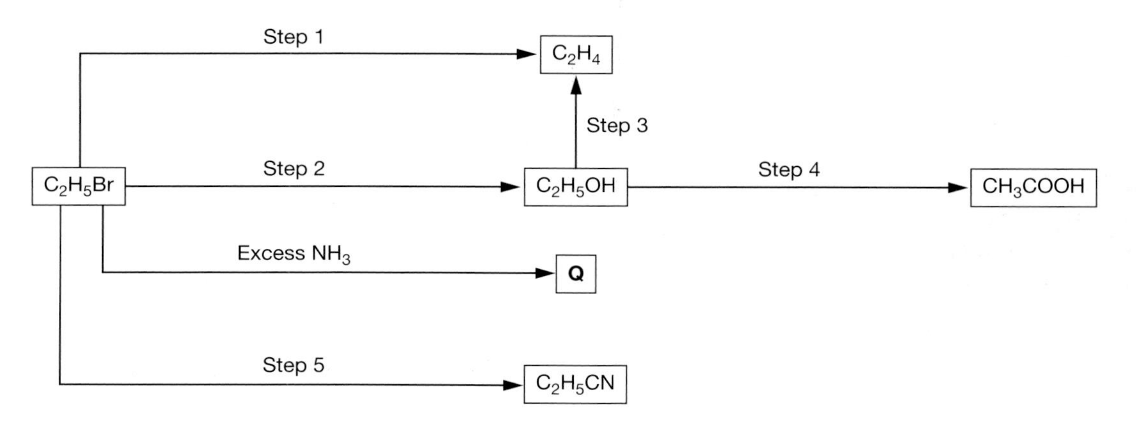

(a) State the reagents and conditions for:

(i) step 1

Reagent: .... Conditions: ....

[3 marks]

(ii) step 2

Reagent: .... Conditions: ....

[2 marks]

(iii) step 3

Reagent: .... Conditions: ....

[2 marks]

(iv) step 4

Reagents: .... Conditions: ....

[2 marks]

(v) step 5 (not OCR candidates)

Reagent: .... Conditions: ....

[3 marks]

(b) What would you observe when carrying out step 4?

....

[2 marks]

(c) Write the full (displayed) structural formula of Q.

[1 mark]

(d) Bromoethane, C_2H_5Br, could be made from ethane.

(i) State the reagent and conditions of this reaction.

Reagent: .... Conditions:..

[2 marks]

(ii) Classify this reaction.

....

[2 marks]

(iii) Why is this not a good method for preparing bromoethane from ethane?

....

....

[1 mark]

[Total 20 marks]

Q2

(a) Compound X is a mixture of two cis/trans (geometric) isomers. Its composition by mass is C = 47.1%, H = 6.5% and Cl = 46.4%.

(i) Calculate the empirical formula of X.

[3 marks]

(ii) The molecular formula is the same as the empirical formula. Draw diagrams showing the two cis/trans isomers of substance X.

[2 marks]

(iii) There are two non-cyclic structural isomers of X. Draw them.

[2 marks]

(b) Halogenoalkanes such as bromoethane, C_2H_5Br, undergo many nucleophilic substitution reactions.

(i) Explain the meaning of the term nucleophile.

....

....

[2 marks]

(ii) Give an example of a nucleophile that will react with bromoethane.

....

[1 mark]

(iii) Explain the meaning of the term substitution.

....

....

[1 mark]

[Total 11 marks]

Q3

(not for Edexcel candidates)

(a) Describe, using curly arrows and polarities in your diagrams, the mechanism for the addition of hydrogen bromide, HBr, to propene, $CH_3CH{=}CH_2$.

[5 marks]

(b) Classify this reaction.

....

[2 marks]

(c) Explain why the product that you have drawn is the major product of this reaction.

....

....

....

[2 marks]

The answers to these questions are on pages 88, 89 and 90.

[Total 9 marks]

Notes: A tick against part of an answer indicates where a mark would be awarded.

Chapter 1 Atomic Structure

Q1 How to score full marks

a) (i) $1s^2, 2s^2, 2p^3$. ✓

(ii) The highest occupied energy level is a p orbital. ✓

(iii) $N(g) \rightarrow N^+(g) + e^-$. ✓ ✓

(iv) N is $1s^2,2s^2, 2p_x^1 2p_y^1\ 2p_z^1$ whereas O is $1s^2,2s^2, 2p_x^2\ 2p_y^1\ 2p_z^1$. ✓

The **repulsion** of to 2 electrons in the $2p_x$ orbital of **oxygen** makes it **easier** to remove one of them. ✓

Although F is $1s^2,2s^2, 2p_x^2 2p_y^2\ 2p_z^1$, the nuclear charge is **two more** than in nitrogen ✓ and there are **no** extra shielding (inner shell) electrons, and so the $2p_x$ electron in fluorine is held more firmly, and so is harder to remove. ✓

b) (i) Mass number is the sum of the number of protons and neutrons ✓ in an atom ✓ (of the isotope).

Relative atomic mass is the average ✓ mass of the naturally occurring isotopes ✓ (taking into account their relative abundance) of the different atoms of the element relative to (or divided by) $\frac{1}{12}$th the mass of a carbon 12 atom. ✓

(ii) % of ^{121}Sb = 60 and % ^{123}Sb = 40. ✓

A_r of Sb = $(121 \times 60 + 123 \times 40) \div 100 = 121.8$ ✓

(iii) The ions are accelerated by an electric field ✓
They are deflected by a magnetic field. ✓
The heavier ions are deflected the least. ✓

Examiner's comments

(a) (ii) Do not say "partially filled p orbital"

(iii) Remember that ionisation energy refers to **gaseous atoms.**

(iv) To gain full marks you have to mention, where appropriate
The change in nuclear charge
The change in shielding (if any)
The repulsion due to two electrons in the same orbital
The size of the atom.

Here the last point is not very important, as the atoms are all of a very similar size.

(b) (i) As the question asked to **distinguish** the two terms, you must make it clear that the mass number refers to a particular isotope of an element, whereas relative atomic mass refers to the natural mixture of the isotopes of an element.

(ii) The four parts to a mass spectrometer are
Ionisation (of a gaseous sample)
Acceleration (by a electric field)
Deflection (by a magnetic field)
Detection (not asked for in this question).

Q2 How to score full marks

(a) **(i)** C and E are isotopes ✓ (B and D are ions not atoms).

(ii) B is a positive ion. ✓

(iii) D is a negative ion. ✓

(iv) A is different from all the others. ✓

(v) A, C and E are atoms. ✓

(b) **(i)** Although number of protons (nuclear charge) increases, ✓ the number of shielding electrons increases by the same amount. ✓

The outer s electron becomes further from the nucleus (atomic radius increases), ✓

and so it is held less firmly, and less energy is required to remove it. ✓

(ii) The nuclear charge increases without ✓ any increase in the number of inner (shielding) electrons. ✓

Therefore the outer 3p electron is held more firmly Al to Si to P. ✓

> **Examiner's comments**
>
> **(a)** **(ii)** B has lost 2 electrons and so formed a 2+ ion
> **(iii)** D has gained 1 electron and so formed a 1− ion
> **(iv)** A has a different atomic (proton) number from all the others
> **(v)** Atoms are neutral and so must have equal numbers of protons and electrons
>
> **(b)** **(i)** Size is the dominant factor within a group, because the increase in nuclear charge is compensated for by the increase in the number of shielding electrons
> **(ii)** The increase in nuclear charge determines the general trend across a period

Chapter 2 Formulae, equations and amount of substance

> **Examiner's comments**
>
> Note how the answers have been set out showing all the working.
>
> This allows partial marks to be awarded if a careless mistake or an error is made.
>
> In **(ii)** the question asked for the molecular formula to be **deduced**. This means that, however simple the calculation, some working has to be given.

Q1 How to score full marks

(a) **(i)** moles of Ca $= \frac{\text{mass}}{A_r} = \frac{0.122}{40} = 0.00305$ mol ✓

moles of HCl = concentration × vol $= 1.23 \times \frac{25}{1000} = 0.03075$ mol ✓

0.00305 mol Ca requires $\frac{2}{1} \times 0.00305 = 0.0061$ mol HCl ✓

$0.0061 < 0.03075$, therefore HCl is in excess. ✓

(ii) moles of $H_2 = 0.00305 \times \frac{1}{1} = 0.00305$ mol ✓

volume of H_2 gas = mol × molar volume = $0.00305 \times 24\,000$

$= 73.2\ \text{cm}^3$. ✓

(b) purity $= \frac{\text{actual volume of hydrogen}}{\text{theoretical volume of hydrogen}} \times 100$ ✓ $= \frac{70.1}{73.2} \times 100 = 95.8\%$ ✓

(c) (i)

	%	÷	*Ar* ✓		÷ smallest ✓	
Na	34.3	÷	23 = 1.49	÷ 1.49 =	1	
C	17.9	÷	12 = 1.49	÷ 1.49 =	1	Empirical formula is $NaCO_2$ ✓
O	47.8	÷	16 = 2.99	÷ 1.49 =	2	

(ii) Formula mass of $NaCO_2 = 23 + 12 + 32 = 67$ ✓

$\frac{134}{67} = 2$. Therefore molecular formula is $Na_2C_2O_4$ ✓

Q2 How to score full marks

Examiner's comments

(a) (ii) Air is 20% oxygen, and so the volume of oxygen has to be multiplied by $\frac{100}{20} = 5$ to get volume of air required to burn the propane.

(b) (ii) CO_2 contains 2 oxygen atoms per molecule, so don't forget to multiply the number of CO_2 molecules by 2 to get the number of oxygen atoms.

(c) Don't forget the stoichiometric ratios as the reactions are not 1:1.

(i) The answer 7.23 is obtained by leaving all the figures of the intermediate answers on the calculator, and then rounding up to 3 significant figures at the end. 7.22 g would get full marks.

(a) (i) $C_3H_8 + 5O_2$ ✓ $\rightarrow 3CO_2 + 4H_2O$ ✓

(ii) as both are gases ✓

volume of O_2 = volume of $C_3H_8 \times \frac{5}{1} = 25 \times 5 = 125\ cm^3$ ✓

volume of air = volume of $O_2 \times 5 = 125 \times 5 = 625\ cm^3$ ✓

(b) (i) moles of $CO_2(g) = \frac{\text{volume}}{\text{molar volume}} = \frac{75}{1000} \div 24$

$= 3.1 \times 10^{-3}$ mol ✓

(ii) number of oxygen atoms = 2 ✓ $\times 6.02 \times 10^{23} \times 3.125 \times 10^{-3}$ ✓ $= 3.8 \times 10^{21}$

(iii) concentration = $\frac{\text{moles}}{\text{volume}} = 3.125 \times \frac{10^{-3}}{1.5} = 2.1 \times 10^{-3}\ mol\ dm^{-3}$ ✓

Q3 How to score full marks

(a) $Al_2O_3 + 6HCl \rightarrow 2AlCl_3 + 3H_2O$ ✓

(b) $4LiNO_3 \rightarrow 2Li_2O + 4NO_2 + O_2$ ✓

(c) $2K + 2H_2O \rightarrow 2KOH + H_2$ ✓

(d) $3Ca(OH)_2 + 2Na_3PO_4$ ✓ $\rightarrow Ca_3(PO_4)_2 + 6NaOH$ ✓

(e) $2Fe^{3+}(aq) + Sn^{2+}(aq) \rightarrow 2Fe^{2+}(aq) + Sn^{4+}(aq)$ ✓

(f) $5I^-(aq) + IO_3^-(aq) + 6H^+(aq)$ ✓ $\rightarrow 3I_2(aq) + 3H_2O(l)$ ✓

Examiner's comments

(a) 2Al in Al_2O_3 requires $2AlCl_3$ on right and hence 6Cl and so 6HCl on left

(d) Either count all the atoms of each element or count OH groups and PO_4 groups separately. $2PO_4$ groups on right, therefore there must be 2 on the left, therefore 6Na's required (in $2Na_3PO_4$), so 6NaOH on right.

(e) This equation must balance for charge. $2x3^+ + 2^+ = 8^+$ on left and $2x2^+ + 4^+ = 8^+$ on right.

(f) 3 oxygen atoms in IO_3^-, therefore $3H_2O$ on right. This means $6H^+$ on left. As the right is neutral there must be $5I^-$ on the left to make it neutral too. 5^- & 1^- & $6^+ = 0$ (neutral).

(c) (i) moles of $Na_2CO_3 = \frac{\text{mass}}{M_r} = \frac{4.56}{106} = 0.0430$ mol ✓

moles of $NaHCO_3$ = moles of $Na_2CO_3 \times \frac{2}{1} = 0.0860$ mol ✓

mass of $NaHCO_3$ = moles $\times M_r = 0.0860 \times 84 = 7.23$ g ✓

(ii) moles H_2SO_4 = concentration × volume $= 5.00 \times \frac{100}{1000} = 0.500$ mol

moles of $NaHCO_3$ = moles $H_2SO_4 \times \frac{2}{1} = 1.00$ mol

mass $NaHCO_3$ = moles $\times M_r = 1.00 \times 84 = 84$ g needed to neutralise spilt acid.

Chapter 3 Structure and bonding

Q1 How to score full marks

(a) The pairs of electrons around the central atom repel each other ✓ and get as far apart as possible. ✓

The lone pair/lone pair repulsion is greater than lone pair/bond pair, which is greater than bond pair/bond pair. ✓

(i)

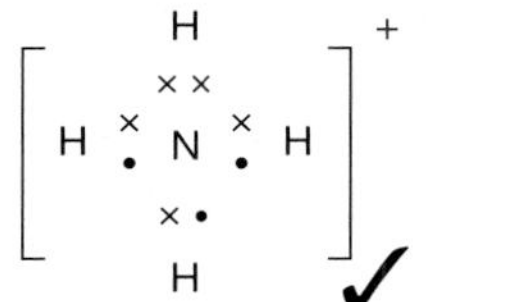

(ii)

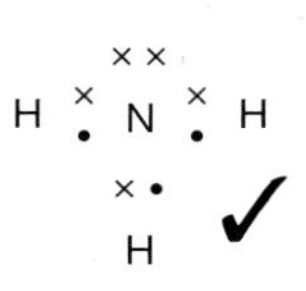

(iii)

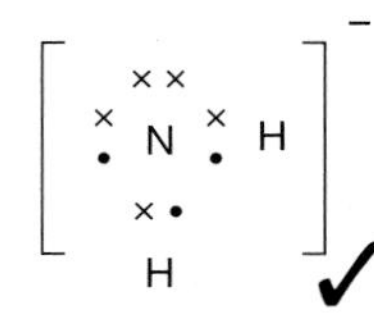

There are 4 bond pairs and no lone pairs, ✓ so the shape is tetrahedral. ✓

There are 3 bond pairs ✓ and 1 lone pair, so the shape is pyramidal. ✓

There are 2 bond pairs and 2 lone pairs, ✓ so it is V-shaped. ✓

(iv) In NH_4^+ there are no lone pairs and so the bond angle is $109\frac{1}{2}°$ ✓

In NH_3 there is one lone pair, which reduces the bond angle (to 107°), and in NH_2^- there are two, reducing the angle even more, so NH_4^+ has the greatest bond angle. ✓

(c) CO_2 is linear ✓ O=C=O ✓ bond angle is 180° ✓

SO_2 is triangular planar ✓ (S̈ double-bonded to two O) ✓ bond angle is 120° ✓

> **Examiner's comments**
>
> **(a)** This is called the (valence shell) electron pair repulsion theory
>
> **(b)** There is one co-ordinate (dative) bond in NH_4^+. In NH_2^- the nitrogen atom has one extra (ionic) electron, and so now has two lone pairs, hence a shape and bond angle similar to those in water.
>
> **(c)** CO_2 has two σ bonds, two π bonds (therefore two areas of negative charge) and **no** lone pairs, and is linear
>
> SO_2 has two σ bonds, two π bonds (making two areas of negative charge) and **one** lone pair, so is triangular planar.

2 How to score full marks

a) The hydrogen bond is an **intermolecular** force ✓ which in water is between a δ+ hydrogen atom in one molecule and a δ- oxygen atom in another molecule. ✓

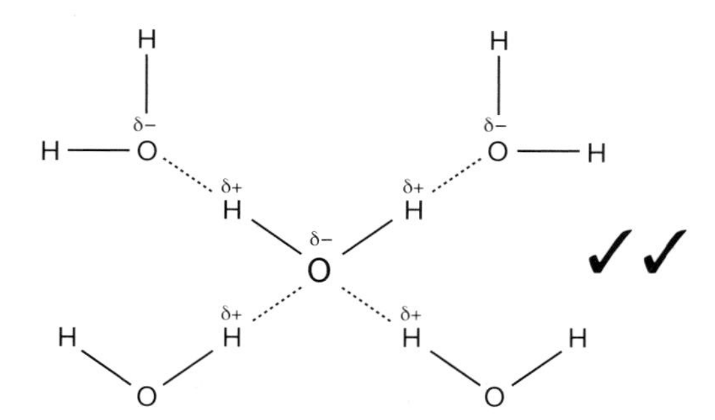

✓✓

b) There are **hydrogen** bonds between HF molecules but not between HCl molecules. ✓ Therefore the **intermolecular** forces are stronger in HF than in HCl, and so HF has a higher boiling point. ✓

In HCl and HBr the main intermolecular forces are instantaneous induced dipole/induced dipole (also called London or dispersion) forces. ✓

HCl has fewer electrons (18) then does HBr (36 electrons), and so the instantaneous induced dipole/induced dipole (dispersion) forces in HCl are weaker than those in HBr, and hence HCl has the lower boiling point. ✓

Examiner's comments

a) You must draw a diagram showing the difference between the covalent bonds and the hydrogen bonds. Note that each oxygen atom in water forms **two** covalent bonds and **two** hydrogen bonds, and each hydrogen **one** covalent and **one** hydrogen bond.

b) Hydrogen bonds only occur between hydrogen and **F**, **O** or **N** atoms.

HCl is more polar than HBr, but the stronger instantaneous induced dipole/induced dipole (dispersion) forces between HBr molecules outweigh the slightly stronger dipole/dipole forces between HCl molecules.

3 How to score full marks

Examiner's comments

a) To decide if a molecule is polar, first decide if the bonds are polar, and then if the polarities cancel because the molecule is symmetrical.

b) Look for the critical differences. The charge difference between Mg^{2+} and Na^{+} is more important than the size difference of the two ions.

c) The more polarising the cation and the more polarisable the anion, the more covalent the bond.

a) Chlorine is more **electronegative** than hydrogen. ✓

Therefore the bonding electrons are pulled closer to chlorine, making the H δ+ and the Cl δ−. ✓

In Cl_2 there is no difference in electronegativity of the two atoms and so the bond is not polar. ✓

In CCl_4 there is polarity in the bonds ✓ but the molecule is symmetrical and so the polarity cancels out making the molecule non-polar. ✓

(b) **(i)** Li^+ ✓
Because it has a smaller ionic radius than Na^+. ✓

(ii) Mg^{2+} ✓
Because it has a charge of 2+ whereas sodium has a charge of only 1^+. ✓

(c) Al^{3+} is very small and is highly charged, so is very polarising. ✓
Cl^- is bigger than F^- and so is more polarisable. ✓
Cl^- becomes so polarised that the bond becomes covalent. ✓

Chapter 4 The Periodic Table and Group II (Group 2)

Q1 How to score full marks

(a) **(i)** The solubility of the group II hydroxides **increases** down the group. ✓

(ii) Name: barium chloride ✓
Ionic equation: $Ba^{2+}(aq) + SO_4^{2-}(aq) \rightarrow BaSO_4(s)$ ✓

(b) **(i)** $CaCO_3 \rightarrow CaO + CO_2$ ✓

(ii) $CaCO_3(s) + 2HCl(aq) \rightarrow CaCl_2(aq) + H_2O(l) + CO_2(g)$ ✓ ✓

(c) **(i)** Strontium, ✓
as it has 1 more shell of electrons than calcium, and so is bigger. ✓

(ii) Magnesium; ✓
the electronegativity decreases down as the group because the atoms become bigger and exert less pull on the bonding electrons. ✓

(d) **(i)** Magnesium: $1s^2, 2s^2 2p^6, 3s^2$ ✓
Phosphorus: $1s^2, 2s^2 2p^6, 3s^2 3p^3$ ✓

(ii) Sulphur forms S_8 molecules whereas chlorine is Cl_2 (or sulphur molecules are much bigger than chlorine molecules). ✓

Thus the instantaneous induced dipole/induced dipole (London or dispersion or van der Waals forces) ✓ are stronger between sulphur molecules than between chlorine molecules. ✓

(iii) Na_2O (or Na_2O_2) or MgO ✓ (not Al_2O_3 as it is amphoteric).

Examiner's comments

(a) (ii) The Cl^- and Na^+ ions are spectator ions, and so must not be included in an ionic equation. State symbols are necessary for this type of reaction as a precipitate is formed from an aqueous solution.

(b) (ii) 1 mark is awarded for a balanced equation, and 1 mark for state symbols. You should know that calcium carbonate is an insoluble solid and that calcium chloride is water soluble.

Q2 How to score full marks

(a) The 1st ionisation energy decreases because the radius increases. ✓
The extra nuclear charge ✓ is compensated for by the equivalent increase in inner (shielding) electrons. ✓

(b) **(i)** $Ca(s) + 2H_2O(l) \rightarrow Ca^{2+}(aq) + 2OH^-(aq) + H_2(g)$ ✓✓

(ii) $Mg(s) + H_2O(g) \rightarrow MgO(s) + H_2(g)$ ✓

(iii) The 1st and 2nd ionisation energies get less ✓ so it gets easier to remove 2 electrons to form a 2^+ ion. ✓

(c) **(i)** Radium reacts very rapidly ✓
to form radium hydroxide and hydrogen
(or the equation $Ra + 2H_2O \rightarrow Ra(OH)_2 + H_2$). ✓

(ii) Insoluble. ✓

(iii) Any value between 218 and 221 pm. ✓

Examiner's comments

(a) There are 3 marks for the question, so 3 points have to be made.

(b) (i) Calcium hydroxide is ionic and (reasonably) soluble so the Ca^{2+} and the OH^- ions have to be written separately. The other substances are not ionic.

(iii) Group II metals form 2+ ions, and so mention must be made of both the 1st and 2nd ionisation energies.

(c) (i) The prediction must state what happens and how quickly.

Chapter 5 Redox and Group VII (Group 7)

Q1 How to score full marks

(a) **(i)** Oxidation number is the charge that an element would have ✓ if all the bonds in the substance were ionic. ✓

(ii) The oxidation number of chlorine in Cl_2 is zero, ✓ in OCl^- is +1 ✓ and in $NaClO_3$ it is +5. ✓

(b) **(i)** $Cl_2 + H_2O \rightleftharpoons HCl + HOCl$ ✓
$Cl_2 + 2OH^- \rightarrow Cl^- + OCl^- + H_2O$ ✓

(ii) The chlorine has been both oxidised and reduced (disproportionated). ✓
Its oxidation number has simultaneously gone up form zero in Cl_2 to +1 in H_2O and OCl^- and down from zero in Cl_2 to –1 in HCL and Cl^-. ✓

(c) **(i)** $FeO_4^- + 8H^+ + 4e^- \rightarrow Fe^{3+} + 4H_2O$ ✓✓

(ii) $Fe^{2+} \rightarrow Fe^{3+} + e^-$ ✓

(iii) $FeO_4^- + 8H^+ + 4Fe^{2+} \rightarrow 5Fe^{3+} + 4H_2O$ ✓✓

Examiner's comments

(a) (ii) The oxidation number of an uncombined element is zero.
In OCl^-, –2 + Cl = –1 and in $NaClO_3$, +1 + Cl + (–6) = 0.

(c) (i) Fe is +7 in FeO_4^-, and iron's oxidation number changes by –4, so $4e^-$ must be on the left. $8H^+$ ions are needed to form $4H_2O$ from the 4 oxygens in FeO_4^-.

(iii) Equation (ii) has to be multiplied by 4, so that both equations have the same number of electrons.

Q2 How to score full marks

(a) Firstly the concentrated sulphuric acid protonates each halide, forming gaseous hydrogen halide. ✓ The equation for the reaction with potassium chloride is

$$KCl + H_2SO_4 \rightarrow HCl(g) + KHSO_4$$

With potassium chloride no further reaction takes place because HCl is a very weak reducing agent. ✓

The observation is steamy fumes (of HCl). ✓

With potassium bromide. Firstly HBr is produced, but some of it oxidised to bromine ✓ and some of the sulphuric acid is reduced to sulphur dioxide. ✓

The observation is a red-brown gas (Br_2). ✓ (The sulphur dioxide is colourless and so not seen. Some HBr fumes may be seen)

With potassium iodide. Firstly HI is produced, which is a strong reducing agent and so reduces the sulphuric acid to solid sulphur ✓

The HI is oxidised to iodine. ✓

The observation is violet vapour (I_2). ✓

(b) **(i)** **Y** is AgCl (silver chloride), ✓
Z is AgI (silver iodide) ✓ ✓ ✓

(ii) The ions in solid **P** are Na^+, Cl^- and I^-. ✓

(iii) $Ag^+(aq) + Cl^-(aq) \rightarrow AgCl(s)$ ✓
$Ag^+(aq) + I^-(aq) \rightarrow AgI(s)$ ✓

Examiner's comments

(a) The extent to which the sulphuric acid oxidises the hydrogen halides depends on the strength of the hydrogen halide as a reducing agent. You must give the names or formulae of the products and what you would see.

(b) W is a mixture of the two silver halides, one is soluble in dilute ammonia and the other insoluble in concentrated ammonia.

Chapter 6 Energetics

Q1 How to score full marks

Examiner's comments

(a) (ii) it is a good idea to give an equation with state symbols as part of an answer to a definition in energetics.

(b) The first two steps of the cycle are the reverse of formation and so equals $-\Delta H_f$.

(c) (i) One O—H bond in water breaks. The H bonds to one carbon atom and the O in the OH bonds to the other carbon atom

(a) **(i)** Standard conditions are a constant pressure of 101 kPa (1 atm), ✓
a stated temperature (usually 298K) ✓
and any solutions at a concentration of 1 mol dm^{-3}. ✓

(ii) It is the enthalpy change when 1 mole of the substance ✓
is formed from its elements under standard conditions. ✓
For ethene it is the enthalpy change for the reaction:
$2C(s) + 2H_2(g) \rightarrow CH_2{=}CH_2(g)$ ✓

(b) $$CH_2{=}CH_2(g) + H_2O(g) \rightarrow CH_3CH_2OH(g)$$ ✓

$$2C(s) + 3H_2(g) + \tfrac{1}{2}O_2(g)$$

$\Delta H_r = -\Delta H_f\text{ (ethene)} - \Delta H_f\text{ (water)} + \Delta H_f\text{ (ethanol)}$ ✓

$= -(+53) - (-242) + (-235) = -46\text{ kJ mol}^{-1}$ ✓

(c) **(i)**

Bonds broken (endothermic)	Bonds made (exothermic)
C═C = +612	C—C = −348
H—O = +463	C—H = −412
	C—O = −360
+1075 kJ ✓	−1120 kJ ✓

$\Delta H_r = +1075 + (-1120) = -45\text{ kJ mol}^{-1}$ ✓

(ii) The bond energies in the data are average values, ✓ so there is a slight difference between the average bond enthalpies and the actual bond enthalpies in these compounds. ✓

Q2 How to score full marks

Examiner's comments

(a) Straight lines need to be drawn for the period up to 3 minutes and from 5 to 10 minutes. These lines are then extrapolated to 3 minutes (when the reaction started) and ΔT measured at the 3-minute point.

ΔH is negative as the reaction is exothermic.

(b) The mass is the sum of the copper sulphate and the water

(c) The difficult part is deducing the cycle. The clue is the ΔH values that you have: the ΔH calculated in part (b), and that given in the data of (c).

(a)

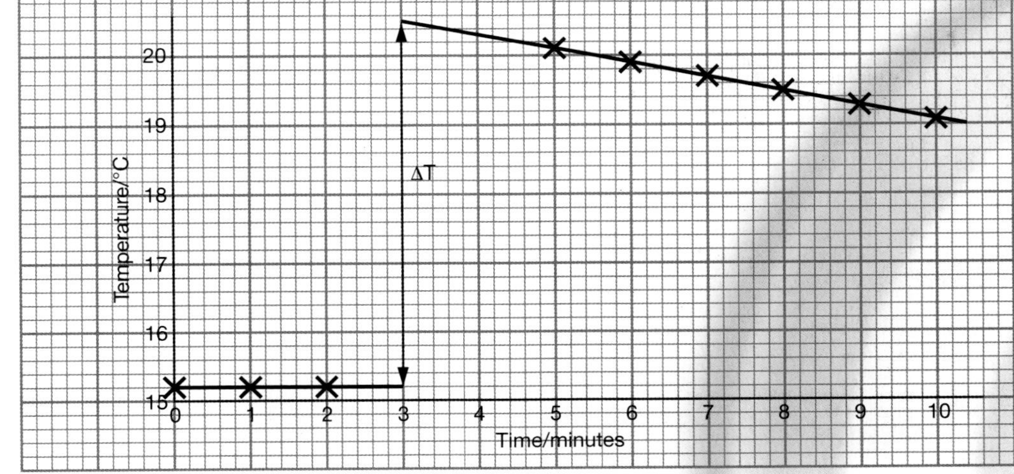

Axes labelled ✓ points correctly drawn ✓ lines drawn ✓ extrapolation to 3 minutes ✓

$\Delta T = 20.5 - 15.2 = 5.3\,°C$. ✓

(b) heat produced = $104.8 \times 4.18 \times 5.3 = 2322\text{ J} = 2.322\text{ kJ}$ ✓

amount of anhydrous copper sulphate = $\dfrac{4.80\text{ g}}{159.5\text{ g mol}^{-1}} = 0.03009\text{ mol}$ ✓

heat produced per mole = $\dfrac{2.322}{0.03009} = 77.2\text{ kJ mol}^{-1}$ ✓

$\Delta H = -77.2\text{ kJ mol}^{-1}$ ✓

(c) 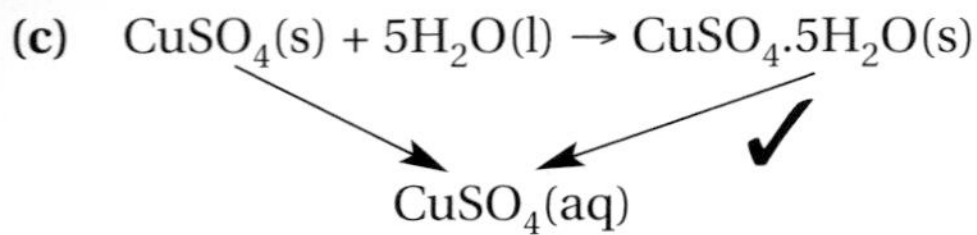

or

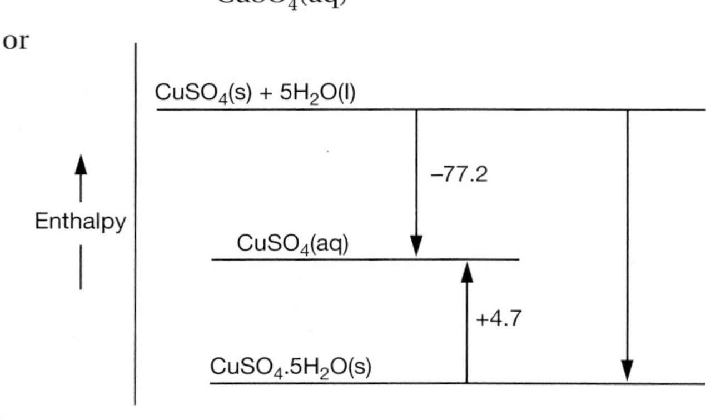

ΔH_r = enthalpy of solution of $CuSO_4(s)$ – enthalpy of solution of $CuSO_4.5H_2O$

$= -77.2 - (+4.7) = -81.9$ kJ mol^{-1}

Chapter 7 Kinetics

Q1 How to score full marks

(a) the rate can be increased by:

an increase of temperature
an increase the concentration of either or both reactants
and the addition of a suitable catalyst.

(b) The flour dust has a very high surface area and is dispersed in the air.

The rate of collision between the oxygen molecules and the surface of the flour particles is very high, and so the reaction is rapid.

In the home the flour is not dispersed in the air.

(c) **(i)** None, there are no reactant gases.

(ii) It slows down the rate because water will lower the concentration of the acid.

(iii) None, because HBr and HCl are both strong acids, and it is the H^+ ions that react with the zinc.

Examiner's comments

(a) This reaction is carried out in solution, so pressure cannot alter the rate of reaction.

(b) To score full marks, you need to say something about both the flour in the mill and in the home.

(c) HCl is fully ionised into $H^+(aq)$ and $Cl^-(aq)$

Q2 How to score full marks

(a) The activation energy is the minimum energy that the molecules must have on collision in order to form the products. ✓✓

(b) (i)

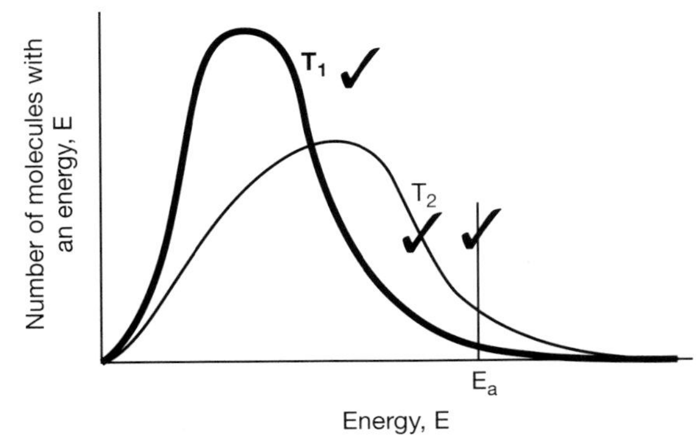

At the higher temperature the molecules have a higher average kinetic energy ✓ and so the area under the curve to the right of the E_a line is greater at T_2 than at T_1. ✓
This means that more molecules will have the activation energy on collision and so will react. ✓
This causes the reaction rate to increase. ✓
In addition there is a slight increase in rate due to the molecules moving faster and so colliding more often. ✓ 8/8

(ii) The rate will increase. ✓
The reactants are gases and so an increase in pressure will result in their being more molecules in a given volume. ✓
This means that the frequency of collision will be increased ✓ 3/3

Examiner's comments

(a) You need to draw two graphs at temperatures T_1 and T_2. These must be clearly labelled with the activation energy drawn in to the right of both peaks and will score three marks. The remaining five marks are for the explanation.

(b) Firstly you must state whether the rate would be unchanged, increased or decreased. This will score 1 mark. There are 3 marks for this part, so you must make 2 more points to score full marks.

Chapter 8 Equilibrium

Q1 How to score full marks

(a)

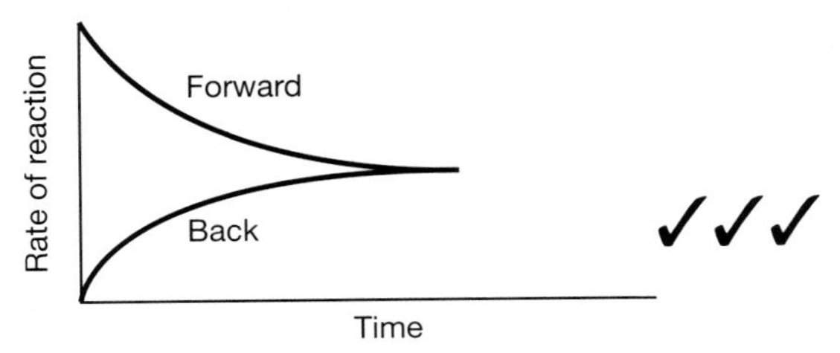

✓✓✓

(b) false ✓
true ✓
false ✓
true ✓

(c) (i) A temperature of 425 °C is chosen because:
as the reaction is exothermic left to right, a higher temperature would shift the position of equilibrium to the left which would result in a lower yield. ✓
a lower temperature would shift the equilibrium to the right causing the theoretical yield to increase, ✓
but the rate of reaction would be too low, and hence the time required to reach equilibrium would be too long. ✓
a compromise temperature of 425 °C is used, which is an economic balance between rate and yield. ✓

(ii) As there are fewer gas molecules on the right hand side, a high pressure would shift the position of equilibrium to the right, thus increasing the yield. ✓

But the yield is high at a pressure of 202kPa, and so there is no need for expensive high pressure. ✓ (The pressure must be above atmospheric to push the chemicals through the plant.)

(iii) A catalyst speeds up the rate and so enables the reaction to take place quickly at a moderate temperature. ✓

It has no effect on the position of equilibrium or the theoretical yield. ✓

Thus it is more economic to run the plant at the same rate at a lower temperature with a catalyst, than without a catalyst when the reaction would require a higher temperature and hence produce a lower yield. ✓

(d) (i) The solution would turn yellow. ✓
The OH^- ions would react with the H^+ ions, thus lowering their concentration. ✓
The system adjusts to replace those H^+ ions by moving to the left. ✓

(ii) The solution would turn orange. ✓
The addition of sulphuric acid increases the concentration of H^+ ions. ✓
The system adjusts to remove them by moving to the right. ✓

Examiner's comments

(a) The rate of the reverse reaction is zero at the beginning. At equilibrium the two rates are equal.

(b) The equilibrium is dynamic and so both reactions are still taking place, but at equal rates.

(c) (i) You must explain the disadvantages of using either a higher temperature or a lower temperature.

(ii) At first sight a high pressure would seem to be a good idea, but high pressures are expensive and so are only used for reactions with low yields (such as the Haber process).

(d) You must state the colour change and then explain it in terms of equilibrium.

Chapter 9 Organic chemistry

Q1 How to score full marks

(a) (i) step 1. Reagent: potassium hydroxide (KOH) ✓
Conditions: heat under reflux ✓ in ethanolic solution. ✓

(ii) step 2. Reagent: aqueous sodium or potassium hydroxide ✓
Conditions: heat under reflux ✓

(iii) step 3 Reagent: concentrated sulphuric or phosphoric(V) acid. ✓
Conditions: heat to 170 °C ✓

Or Reagent: aluminium oxide, Al_2O_3, as catalyst
Conditions: pass ethanol vapour over hot aluminium oxide

(iv) step 4 Reagent: potassium dichromate(VI) + dilute sulphuric acid ✓
Conditions: heat under reflux ✓

(v) step 5 Reagent: potassium cyanide ✓
Conditions: heat under reflux ✓ in an aqueous ethanol solvent ✓

(b) The solution would go from orange ✓ to green. ✓

(c)

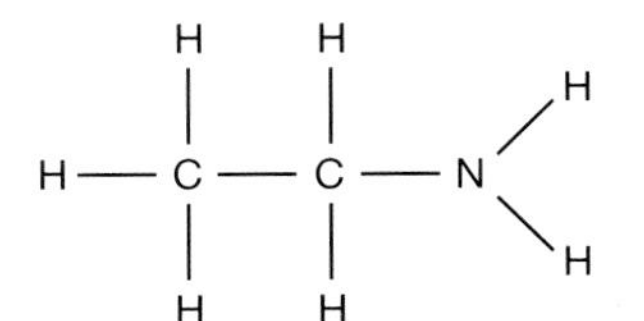

or

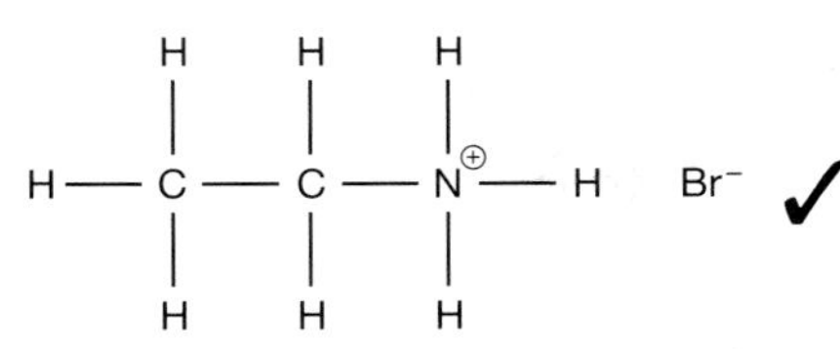

✓

(d) (i) Reagent: bromine ✓ Conditions: UV (or sun) light ✓

(ii) Free radical ✓ substitution ✓

(iii) Some $C_2H_4Br_2$ and further substitution products would also be produced. ✓

Examiner's comments

(a) Step 1 is elimination and so an ethanolic solution must be used, whereas step 2 is hydrolysis and aqueous alkali must be used.
Step 3 is dehydration and so the reagent is concentrated sulphuric acid or aluminium oxide.
Step 4 is oxidation. If heated under reflux, an acid is produced (but if the product is allowed to distill off as it is made, the aldehyde will be obtained).

(b) Do not forget to give the colour before and after reaction.

(c) Both the free amine and its salt can be given as answers.

Q2 How to score full marks

Examiner's comments

(a) (i) Do not round down to a whole number after dividing by the relative atomic mass.

(ii) The question told you that **X** has cis/trans isomerism, and so the compound must contain a C=C group with two different groups on each carbon atom.

(iii) In these compounds, you must have two hydrogen atoms on one of the carbon atoms.

(b) (i) Remember that nucleophiles must be defined in terms of a lone pair.

(iii) Do not use the word substitutes in your definition of substitution.

(a) (i)

	%	$\div A_r$ ✓		$\div 1.31$ ✓	empirical formula
carbon	47.1	$\div 12$	$= 3.93$	$= 3$	C_3H_5Cl ✓
hydrogen	6.5	$\div 1$	$= 6.5$	$= 5$	
chlorine	46.4	$\div 35.5$	$= 1.31$	$= 1$	

(ii)

Cl(H)C=C(CH$_3$)(H) ✓ Cl(H)C=C(H)(CH$_3$) ✓

(iii)

H$_2$C=C(H)(CH$_2$Cl) ✓ H$_2$C=C(Cl)(CH$_3$) ✓

(b) **(i)** A nucleophile is a substance which uses its lone pair of electrons to form a covalent bond. ✓

It attacks electron deficient centres. ✓

(ii) Any one of OH^-, NH_3 or CN^-. ✓

(iii) When one atom or group replaces another in a molecule. ✓

Q3 How to score full marks

(a)

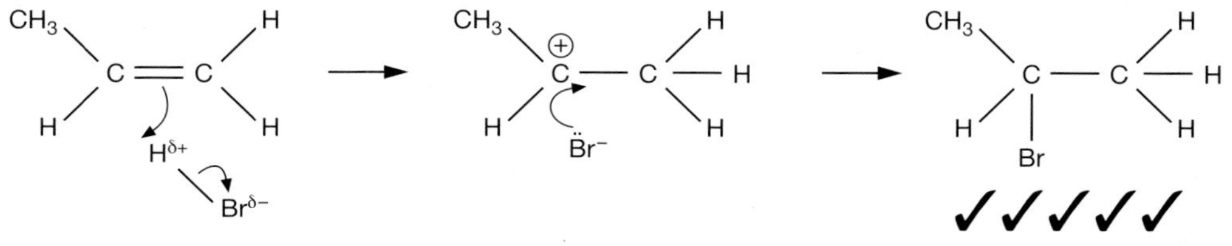

CH_3 H C=C H H $H^{\delta+}$ $Br^{\delta-}$ → CH_3 ⊕ C—C—H H H H Br^- → CH_3 H C—C—H H H Br

✓✓✓✓✓

(b) Electrophilic addition ✓✓

(c) The secondary carbocation ($CH_3CH^+CH_3$) formed when the hydrogen bonds to the end (terminal) carbon atom ✓ is more stable than the primary carbocation ($CH_3CH_2CH_2^+$) which would be formed if the hydrogen bonded to the middle carbon in step 1 of the mechanism. ✓

Examiner's comments

(a) Remember that a curly arrow represents the **movement of a pair** of electrons. One arrow starts from the π bond and goes to form a bond with the H atom in HBr. Another goes from the σ HBr bond to the Br atom. The intermediate has a + charge, and finally the curly arrow starts from the lone pair (not the –) on the bromide ion and goes to form a bond with the + carbon atom.

(b) It is the intermediate secondary carbocation not the product that is more stable

Practice AS Level Test Paper

Time allowed: $1\frac{1}{2}$ hours

1 (a) State the type of bonding in:

(i) sodium chloride

type ....

(ii) silicon tetrafluoride

type .... [2 marks]

(ii) Explain why aluminium fluoride is ionic and boils at 1270 °C whereas anhydrous aluminium bromide is covalent and boils at 265 °C.

....

.... [6 marks]

(b) Silicon is an element in the *p*-block of the Periodic Table.

(i) Using 1s, 2s, 2p … notation, write the electronic structure of silicon.

.... [2 marks]

(ii) State the number of unpaired electrons in one atom of silicon.

.... [1 mark]

(iii) Silicon has three naturally occurring isotopes, ^{28}Si, ^{29}Si and ^{30}Si. Explain why its relative atomic mass is 28.1 not 29.0.

.... [2 marks]

2 (a) 25.0 cm^3 of a 0.111 mol dm^{-3} solution of sodium carbonate, Na_2CO_3, was pipetted into a conical flask and two drops of phenol phthalein indicator added. This solution required 21.3 cm^3 of a 0.130 mol dm^{-3} solution of hydrochloric acid, HCl, for neutralization.

(i) Calculate the moles of sodium carbonate taken.

.... [2 marks]

(ii) Calculate the moles of hydrochloric acid needed.

.... [2 marks]

(iii) Hence, calculate the mole ratio of sodium carbonate to hydrochloric acid.

.... [1 mark]

(iv) Based on these results, suggest the equation for the reaction.

.... [2 marks]

(b) Hydrogen chloride gas, HCl(g), does not react with concentrated sulphuric acid, but when hydrogen bromide gas is bubbled into concentrated sulphuric acid, a brown gas is produced.

(i) Write the equation for the reaction between hydrogen bromide and sulphuric acid.

.... [2 marks]

(ii) Explain this reaction in terms of redox.

.... [4 marks]

(iii) Why is there no reaction between hydrogen chloride and sulphuric acid?

.... [1 mark]

3 2-bromopropane can be hydrolysed by warming with aqueous sodium hydroxide solution. They react according to the equation:

$$CH_3CHBrCH_3 + NaOH \rightarrow CH_3CH(OH)CH_3 + NaBr$$

2-bromopropane has a boiling point of 59 °C and is slightly soluble in water. It is not flammable Propan-2-ol has a boiling point of 82 °C and is very soluble in water. It is very flammable.

(a) When 4.92 g of 2-bromopropane was warmed at 40 °C with an excess of a 1.0 mol dm^{-3} aqueous solution of sodium hydroxide for 20 minutes, 1.80 g of propan-2-ol was produced.

(i) Calculate the percentage yield.

....

.... [4 marks]

(ii) State two ways by which the rate of this reaction could be increased.

.... [2 marks]

(iii) Give two reasons why you would use a flask fitted with a reflux condenser rather than an open beaker for this reaction.

....

.... [2 marks]

(b) Explain why propan-2-ol:

(i) Has a higher boiling point than 2-bromopropane.

.... [4 marks]

(ii) Is more soluble in water than 2-bromopropane.

.... [2 marks]

4 (a) (i) Define standard enthalpy of combustion.

.... [3 marks]

(ii) Illustrate your answer with an equation that represents the standard enthalpy of combustion of ethane, $C_2H_6(g)$.

.... [2 marks]

(b) The standard enthalpies of combustion of ethene, $C_2H_4(g)$, ethane, $C_2H_6(g)$ and hydrogen, $H_2(g)$ are -1409 kJ mol^{-1}, -1560 kJ mol^{-1} and -286 kJ mol^{-1}. Draw a Hess's Law diagram and use it to show that for the reaction:

$$C_2H_4(g) + H_2(g) \rightarrow C_2H_6(g)$$

the value of ΔH equals -135 kJ mol^{-1}.

[4 marks]

(c) Explain what factors determine the bond angles in a covalent compound.

.... [3 marks]

(d) Hence predict the CCH bond angle in:

(i) ethane.

.... [3 marks]

(ii) ethene.

.... [3 marks]

5 (a) Both propane and propene react with bromine, but the conditions and the type of reaction are quite different, as propane reacts by free radical substitution and propene by electrophilic addition. Explain the meaning of the terms:

(i) free radical substitution.

.... [2 marks]

(ii) electrophilic addition.

.... [4 marks]

(b) Propene is manufactured in large quantities by the cracking of a fraction from the distillation of crude oil. Its main use is in the manufacture of poly(propene). Draw the structure of a fragment of poly(propene) showing two repeat units.

.... [2 marks]

(c) The addition of bromine to propene produces $CH_3CHBrCH_2Br$. Draw the three other isomers of this product.

[3 marks]

(d) A sample of a halogenoalkane was heated under reflux with aqueous sodium hydroxide for five minutes. Describe tests that you would do to the resulting solution to show that the original halogenoalkane was a bromo-compound.

....

.... [4 marks]

6 (a) (i) Write the equation (with state symbols) which represents the first ionization energy of nitrogen.

.... [2 marks]

(ii) Explain why there is a big jump in values between the fifth and sixth ionization energies of nitrogen.

.... [3 marks]

(b) Nitrogen dioxide, NO_2, is a dark brown gas and is in equilibrium with dinitrogen tetroxide, N_2O_4, a pale almost colourless gas.

$$2NO_2(g) \rightleftharpoons N_2O_4(g)$$

(i) The colour changes from pale to dark brown as the mixture is heated. Deduce whether the reaction as written is exothermic or endothermic.

.... [3 marks]

(ii) Explain what would happen to the colour of the mixture when the pressure is increased at constant temperature.

.... [3 marks]

AS Trial Paper answers

Q1 How to score full marks

(a) (i) sodium chloride is ionic. ✓

silicon tetrafluoride is covalent. ✓

(ii) aluminium fluoride is ionic because the fluoride ion is so small✓ that it is not polarized✓ by the Al^{3+} ion. It boils at a high temperature because the force between the ions is very strong. ✓

Aluminium bromide is covalent because the much larger✓ bromide ion is polarized by the aluminium ion which has so a high charge density✓ that the bond becomes covalent. It has a much lower boiling temperature because the van der Waals forces✓ between the molecules are not very strong.

(b) (i) Either $1s^2, 2s^2\ 2p^6$✓$3s^2\ 3p^2$✓

or [Ne]✓ $3s^2\ 3p^2$✓

(ii) It has two✓unpaired electrons.

(iii) The % of the ^{29}Si and ^{30}Si isotopes is very small✓, therefore the average✓mass comes out to 28.1.

Total 13 marks

Q2 How to score full marks

(a) (i) the amount of sodium carbonate

$= 0.111 \times \frac{25.0}{1000}$ ✓ $= 0.002775$ mol ✓

(ii) the amount of hydrochloric acid

$= 0.130 \times \frac{21.3}{1000}$ ✓ $= 0.002769$ mol ✓

(iii) the ratio is 1:1 ✓

(iv) $Na_2CO_3 + HCl \rightarrow NaHCO_3$✓$+ NaCl$✓

(b) (i) $2HBr + H_2SO_4 \rightarrow Br_2$✓$+ SO_2 + 2H_2O$

balanced equation✓

(ii) The oxidation number of bromine has gone up from –1 to 0✓and so it has been oxidized✓. The oxidation number of sulphur has gone down from +6 to +4✓ and so it has been reduced.✓

(iii) HCl is not a strong enough reducing agent✓ to be reduced by the concentrated sulphuric acid.

Total 14 marks

Q3 How to score full marks

(a) (i) amount of 2-bromopropane

$= \frac{4.92\text{ g}}{123.0\text{ g mol}^{-1}} = 0.0400$ mol✓

theoretical amount of propan-2-ol
$= 0.0400$ mol✓

theoretical mass of propan-2-ol

$= 0.0400\text{ mol} \times 60\text{ g mol}^{-1} = 2.40$ g✓

% yield $= 100 \times \frac{1.80}{2.40} = 75.0$ %✓

(ii) Increase the temperature✓ (but not above 59 °C).

Increase the concentration of the sodium hydroxide solution.✓

(iii) To prevent the 2-bromopropane (or the organic compounds) from boiling off.✓

To prevent the propan-2-ol from catching fire.✓

(b) (i) In propan-2-ol there is hydrogen bonding✓ between the δ+ hydrogen of the OH group in one molecule and the δ- oxygen in another molecule✓, but the intermolecular forces in 2-bromopropane are only the weaker✓ dipole/dipole and induced dipole/dipole (dispersion) forces.✓

(ii) Propan-2-ol can form hydrogen bonds with the water✓ whereas 2-bromopropane cannot.✓

Total 14 marks

Q4 How to score full marks

(a) (i) It is the enthalpy change when 1 mol✓ of a substance is completely✓ burnt in oxygen under standard conditions (1 atm pressure and a stated temperature, usually 298K)✓

(ii) $C_2H_6(g) + 3\frac{1}{2}\ O_2(g) \rightarrow 2CO_2(g) + 3H_2O(l)$

Balanced equation for 1 mol✓ state symbols✓ (note that water is liquid)

(b)

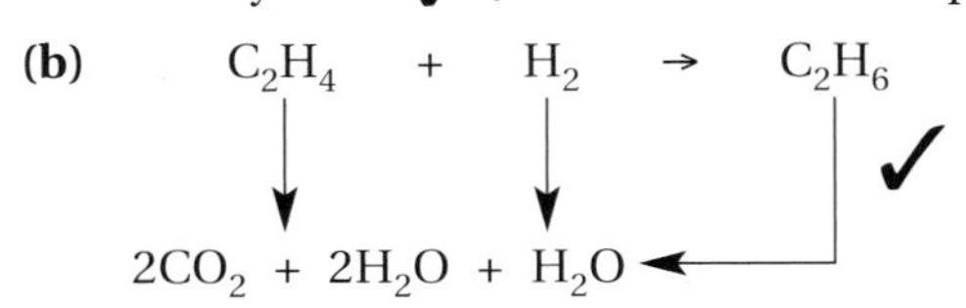

$\Delta H_r = \Delta H_c$ (ethene) + ΔH_c (hydrogen) – ΔH_c (ethane)✓ (or from labelled Hess's law diagram)

$= (-1409) + (-286) - (-1560)$✓$= -135\text{ kJ mol}^{-1}$✓

(c) It is determined by the number of pairs of electrons around the atom✓ which repel to get as far apart as possible✓, but lone pair/lone pair repulsion is greater than lone pair/bond pair which in turn is greater than bond pair/bond pair.✓

(d) (i) Ethane: 4 separate bond pairs✓ arrangement is tetrahedral✓ angle $109\frac{1}{2}°$.✓

(ii) Ethene: 3 separate bond pairs✓ arrangement square planar✓ angle 120°.✓

Total 18 marks

Q5 How to score full marks

(a) (i) A free radical is an atom with an **unpaired** electron.✓

Substitution is when one species (atom, ion or group) replaces another.✓

(ii) An electrophile is a species which will accept a pair of electrons✓ to form a covalent bond✓. It attacks electron rich (δ-) sites.✓

Addition is when two substances form a single compound.✓

(b) $—CH(CH_3)—CH_2—CH(CH_3)—CH_2—$
carbon skeleton✓ continuation bonds✓

(c) $CHBr_2CH_2CH_3$✓ $CH_2BrCH_2CH_2Br$✓
$CH_3CBr_2CH_3$✓

(d)

Either	**or**
acidify with dilute nitric acid✓	acidify with dilute nitric acid✓
add silver nitrate solution✓	add chlorine water✓
bromo compounds give a cream (pale yellow) ppt✓ which is insoluble in dilute ammonia but soluble in concentrated ammonia✓	shake with hexane✓ (or another immiscible organic solvent) bromo compounds give a red/brown colour to the hexane layer✓

Total = 15 marks

Q6 How to score full marks

(a) (i) $N(g) \rightarrow N^+(g) + e^-$ equation✓ state symbols.✓

(ii) The sixth electron comes from an inner shell (or 1st orbit)✓ so is nearer to the nucleus✓ and is not shielded by inner electrons, so is held on more firmly.✓

(b) (i) The equilibrium shifts to the left (towards dark brown NO_2).✓

An increase in temperature shifts the equilibrium in the endothermic direction.✓

So the reaction, as written, is exothermic.✓

(ii) The equilibrium moves to the side with the fewer gas molecules✓ that is to the right,✓ therefore the gas gets paler.✓

Total = 11 mar

TOTAL MARK FOR THE PAPER = 8